Health, Technology and Society

Medicine, health care, and the wider social meaning and management of health are undergoing major changes. In part this reflects developments in science and technology, which enable new forms of diagnosis, treatment and delivery of health care. It also reflects changes in the locus of care and the social management of health. Locating technical developments in wider socio-economic and political processes, each book in the series discusses and critiques recent developments in health technologies in specific areas, drawing on a range of analyses provided by the social sciences. Some have a more theoretical focus, some a more applied focus but all draw on recent research by the authors. The series also looks toward the medium term in anticipating the likely configurations of health in advanced industrial society and does so comparatively, through exploring the globalization and internationalization of health.

More information about this series at
http://www.palgrave.com/gp/series/14875

Dana Mahr · Martina von Arx
Editors

De-Sequencing

Identity Work with Genes

palgrave
macmillan

Editors
Dana Mahr
Section de Biologie
University of Geneva
Geneva, Switzerland

Martina von Arx
Section de Biologie
University of Geneva
Geneva, Switzerland

Health, Technology and Society
ISBN 978-981-15-7727-7 ISBN 978-981-15-7728-4 (eBook)
https://doi.org/10.1007/978-981-15-7728-4

This Palgrave Macmillan imprint is published by the registered company Springer Nature Singapore Pte Ltd.
The registered company address is: 152 Beach Road, #21-01/04 Gateway East, Singapore 189721, Singapore

Series Editors' Preface

Medicine, health care, and the wider social meanings and management of health are continually in the process of change. While the 'birth of the clinic' heralded the process through which health and illness became increasingly subject to the surveillance of medicine, for example, surveillance has become more complex, sophisticated, and targeted—as seen in the search for 'precision medicine' and now 'precision public health'. Both surveillance and health itself emerge as more provisional, uncertain, and risk-laden as a consequence, and we might also ask what now constitutes 'the clinic', how meaningful a concept of a clinic ultimately is, and where else might we now find (or not find) healthcare spaces and interventions.

Ongoing developments in science and technology are helping to enable and propel new forms of diagnosis, treatment, and the delivery of health care. In many contexts, these innovations both reflect and further contribute to changes in the locus of care and burden of responsibility for health. Genetics, informatics, imaging—to name but a few—are redefining collective and individual understandings of the body, health, and disease. At the same time, long established and even ostensibly mundane technologies and techniques can generate ripples in local discourse and practices as ideas about the nature and focus of healthcare shift in response to global debates about, for instance, One Health and Planetary Health.

The very technologies that (re)define health are also the means through which the individualisation of health care can occur—through, for instance, digital health, diagnostic tests, and the commodification of restorative tissue. This individualisation of health is both culturally-derived and state-sponsored, as exemplified by the promotion of 'self-care'. These shifts are simultaneously welcomed and contested by professionals, patients, and wider publics. Hence they at once signal and instantiate wider societal ambivalences and divisions

This Series explores these processes within and beyond the conventional domain of 'the clinic', and asks whether they amount to a qualitative shift in the social ordering and value of medicine and health. Locating technical use and developments in wider socio-economic and political processes, each book discusses and critiques the dynamics between health, technology, and society through a variety of specific cases, and drawing on a range of analyses provided by the social sciences.

The Series has already published more than twenty books that have explored many of these issues, drawing on novel, critical, and deeply informed research undertaken by their authors. In doing so, the books have shown how the boundaries between the three core dimensions that underpin the whole Series—health, technology, and society—are changing in fundamental ways.

In 'De-Sequencing', the editors and contributors focus on an area that has galvanised so much of the scholarship that the HTS series draws from and extends: genetics. This interdisciplinary collection—which draws together researchers from the social sciences and humanities across a range of countries—considers the movement of genetic information between a number of sites, both material and discursive. In doing so, it illustrates the complexities of, and transformations that can happen through, diverse forms of medi(atis)ation, and the ways in which personal identities shape and are shaped by responses to genetic information. Accordingly, 'De-Sequencing' challenges any claim that persons 'are' their genetic sequences, and vividly demonstrates the traffic between health, technologies, and societies.

London, UK Rebecca Lynch
Edinburgh, UK Martyn Pickersgill

Contents

Notes on Contributors

Sarah Ambrosi is a writer and artist living in Berlin, Germany.

Gaia Barazzetti is a philosopher and medical ethicist at The Collaboratory, a participatory, collaboratory and action-research unit at the University of Lausanne, Switzerland.

Claudine Burton-Jeangros is Professor and Sociologist of Health at the Institute of Sociological research at the University of Geneva, Switzerland.

Luca Chiapperino is a philosopher and senior researcher at the Science and Technology Studies Laboratory of the University of Lausanne, Switzerland.

Sara Green is Assistant Professor in the Section for History and Philosophy of Science, Department of Science Education, University of Copenhagen, Denmark.

Yael Hashiloni-Dolev is Professor of Sociology, Department of Sociology and Anthropology, Ben-Gurion University of Negev, Be'er-Sheva, Israel.

Stephan Kopsieker works for the press department of the scientific publishers Ferdinand Schöningh Verlag and Wilhelm Fink Verlag, Germany.

Maria Kronfeldner is a professor in the Department of Philosophy at the Central European University, Hungary.

Grischa Lichtenberger is an electronic music producer, and visual and installation artist living in Berlin, Germany.

Dana Mahr is a sociologist, historian and senior researcher within the ERC/SNF research group 'Rethinking Public Participation and Science' at the University of Geneva, Switzerland.

Kate O'Riordan is Professor of Digital Culture, and Head of School of Media, Film and Music at the University of Sussex, United Kingdom.

Francesco Panese is a professor at the Institute of Social Sciences, and the Sciences and Technologies Studies Laboratory at the University of Lausanne, Switzerland.

Barbara Prainsack is a professor at the Department of Political Science at the University of Vienna, and at the Department of Global Health & Social Medicine at King's College London.

Christoph Rehmann-Sutter is Professor of Theory and Ethics of Biosciences at the Institute for History of Medicine and Science Studies, University of Lübeck, Germany.

Barbara Katz Rothmann is Professor of Sociology at the Graduate Center of the City University of New York, USA.

Silja Samerski is Professor of Social Work and Health at the University of Applied Sciences Emden/Leer, Germany.

Henrik Vogt is a medical doctor who is also educated in History, Journalism and Philosophy. His Ph.D. entitled "Systems medicine as a theoretical framework for primary care medicine—a critical analysis" looked at systems medicine (an aspect of precision medicine or personalized medicine), its theoretical underpinnings and the credibility of its promises. He is currently working at the Centre for Medical Ethics at

The University of Oslo, Norway and is there affiliated to the Hybrid Technology Hub. He is working on epistemological issues in personalized (precision) medicine.

Martina von Arx is a Ph.D. student within the research group 'Development of Personalized Health in Switzerland: Social Sciences Perspectives' at the University of Geneva, Switzerland.

List of Figures

1

Introduction

Dana Mahr, Christoph Rehmann-Sutter, and Martina von Arx

At some point after the completion of the Human Genome Project, when one version of the human DNA sequence had been established, an interesting neologism emerged both in scientific discourse and colloquial language: 'to sequence' a species, a population, a person. Rice has been sequenced, tomatoes, silkworms, and people living in Iceland. It was clear to us all that the process of sequencing referred to the *genome*, not to the species, to the population, or to the person. To sequence *somebody*, has always been an abbreviation, an economy of words—or perhaps

D. Mahr (✉) · M. von Arx
Section de Biologie, University of Geneva, Genève, Switzerland
e-mail: dana.mahr@unige.ch

M. von Arx
e-mail: martina.vonarx@unige.ch

C. Rehmann-Sutter
Institut für Medizingeschichte und Wissenschaftsforschung, Universität zu Lübeck, Lübeck, Germany
e-mail: rehmann@imgwf.uni-lubeck.de

D. Mahr and M. von Arx (eds.), *De-Sequencing*, Health, Technology and Society,
https://doi.org/10.1007/978-981-15-7728-4_1

a slip of the tongue. This book starts from the premise that genetic information contributes in many ways to the construction of identity. The genomic sequence adds up to the identity-forming material from which life is made. Having a closer look at how genetic information is used in conversation will help us to understand its contribution to identity work. Identity cannot adopt the form of a sequence, therefore the genome must be *de*-sequenced by human language to render it interpretable and meaningful in a social context.

The following examples show how the verb 'to sequence' is used, as it would be intelligible to all: Yang Huanming, founder of the Beijing Genomics Institute, one of the world's leading sequencing institutes, declared in the documentary 'DNA Dreams':

> We have initiated the sequencing of 250 small families with parents and one child. We have [also] initiated the sequencing of 5000 twins. (van der Haak 2013)

Amy Dockser Marcus wrote, in an article in *The Wall Street Journal* entitled 'Genetic Testing Leaves More Patients Living in Limbo':

> The expanding use of genetic testing is having an unforeseen consequence: More people are being told they have genes for potentially fatal diseases but don't show any symptoms. As more humans are sequenced and more mutations are [found] … in healthy people who don't have what the mutation is supposed to predict [fear is growing]. (Marcus 2013)

The popular science blogger and educational influencer *Caydunbar*, also an organic chemist at Colorado State University, wrote:

> When people learn that I have been sequenced as part of George Church's Personal Genome Project (PGP) they often say, 'Wow – I'd like to be sequenced too!' My first response to them is, 'Why – are you sick?' (Cayundbar 2013)

John Lauerman, a Bloomberg journalist who had his DNA sequenced, recalls the events in similar words:

> Before I was sequenced, I didn't know I had this clotting disorder. Now I do, and since I began taking baby aspirin, I haven't had a single scotoma. I could be saving myself from all kinds of serious potential illnesses, maybe a stroke or a heart attack. (Lauerman 2012)

In such statements from and about 'sequenced people', as readers, we automatically reintroduce the skipped 'genome of…' into the sentences to make them complete. But is there more to say about this shorthand? Why does this kind of *sequencing language* sound so familiar to many of us?

A DNA molecule can be sequenced because it consists of a linear array of four different nucleotides denoted adenine (A), thymine (T), guanine (G), and cytosine (C). Their sequence matters in roughly the same way as the sequence of letters matters for the words you are reading here. An organism, however, as a dynamic, complex, and corporeal being, cannot in any conceivable way be sequenced. How could a person be sequenced? A deeper question emerges here: How can health and disease, lives, biographies, families, and their narratives be meaningfully related to genetic sequences? There is more than just a formal incompatibility between a life and a sequence. The ambiguity of the term 'sequence' is of a spatial and a temporal dimension. The DNA nucleotides are aligned, but life itself is not executed in a linear form of genetic instructions picked up one after another from the genome. Life is not the temporal version of a row of genes. The difference between life and a genetic sequence is categorical.

Therefore, we need to ask how people establish such a relationship across the categorical differences when they narrate biographies while referring to their genes? What happens in our heads when the verb 'to sequence' is used in such a way? On which levels of biological meanings inherent in living creatures is a rice plant, a silkworm, or a human person sequenceable? And then, what kind of thinking is emerging culturally along with the burgeoning sequencing industries worldwide? In other words, into what kind of thing must species, organisms, persons, and lives be transformed, in order to *become* sequenceable?

Such questions are important and must be addressed carefully. Since the 1990s, much ink has been spilled in discussions about the impact of genetics and genomics on both society and our individual lives. In particular, sociologists, philosophers, researchers with feminist backgrounds, and scholars from the field of disability studies have explored the various ways in which the rise of genetic knowledge—as soon as it transcends the institutional boundaries of the biosciences—*might* reshape our understanding of health and illness, race and gender, moral and ethics, belonging and heritage, as well as work and recreation. The first wave of such research was dominated by scholars revisiting Michel Foucault's writings about biopower and governmentality against the backdrop of genetics. By coining concepts such as *geneticisation* or redefining pre-existing terms such as *biosociality* or *biocapital*, Abby Lippmann (1991, 1999), Dorothy Nelkin and Susan M. Lindee (2004 [1994]), Sarah Gibbon and Carlos Novas (2007) and others, discussed the ongoing developments in microbiology predominantly at the macro-level of society itself. Their retrospective, and rather speculative, studies reflected a societal discourse in which the practical impact of genetics and genomics on society was (at least partly) foreseeable but not yet a ubiquitous reality. Empirical explorations of genetics and genomics in everyday life and on the level of individual sense-making therefore remained basically a desideratum.

One of the rare exceptions was Monica Konrad's anthropological study of 2005 *Narrating the New Predictive Genetics: Ethics, Ethnography, and Science* (Konrad 2005). In the early 2000s, genome sequencing was still an expensive, cutting-edge technology, limited predominantly to prestigious research institutions. Then the rapid development of sequencing technologies during the 2010s, and big data technologies, made the production of genomic data not only faster but also far less expensive and therefore increasingly accessible for clinics, smaller universities, private enterprises, and ultimately also individuals. As a result, a second wave of studies focused mainly on practical ethical problems—for example, whether researchers have an obligation to report incidental additional findings of genome studies to study participants (Wolf et al. 2008; Green et al. 2013; Vayena and Tasioulas 2013; Mahr and Rehmann-Sutter 2015). In recent years, a third wave of studies has become prominent, examining individual identity and genetics from a predominantly empirical point of view. The central question of this strand of research on the micro- and meso-levels of society asks: 'Am I my genes?' (Klitzman 2012). It has united a wide variety of theories, methods, and disciplinary perspectives, from phenomenological and autobiographical or fictional narrative approaches (e.g. Cadwalladar 2014) to qualitative social science studies exploring the emergence of genetic identities in terms of race, ability, gender, citizenship, or consumerism (e.g. Scully 2008; Munsterhjelm 2014; Rehmann-Sutter and Mahr 2016). This third wave of social and philosophical studies on genetics and genomics tracked the different layers of meaning making from the micro-molecular entity in a laboratory up to individuals who decided to adopt a specific lifestyle after receiving their test results from a commercial genetic testing company.

This book adds to the third wave. It has a twofold purpose. On the one hand it seeks to identify and explore various transformations, translations, and meanings of genetic information from laboratories into policies, from clinics to patients, from enterprises to customers, and from graphs and statistics on computer screens to individual identities. On the other hand, it aims at taking up an opposite position to the reductionist paradigm which equates creatures with their genomic sequence. *Sequencist reductionism* deserves systematic and careful counteraction.

DNA sequences return to human lives in forms other than sequences. If this is true, we need to explore how this happens and who is involved in the related activities of the necessary backward translation. The path from rough genetic information to a message relevant in real life involves work. For example, if a genetic sequence reveals the origin of ancestors, the construction of one's own identity would possibly be impacted by the result. This work of translation can be called identity work. There are many ways in which human beings create and socially co-create concepts and descriptions about themselves in reference to their environment, experiences, values, and beliefs (Snow and Anderson 1987). Expressions of identity are not limited to a state of mind but are also incorporated in patterns of behaviour, everyday practices, and are often embedded in social relations. The concept of identity work underwent various theoretical differentiations in the 2000s. W. E. Douglas Creed and Maureen A. Scully explored, for example, the performance-based elements of identity work by incorporating Erwin Goffman's understanding of social encounters and interactivity (Creed and Scully 2000).

The study of identity work is therefore not only concerned with how individuals perform and categorise themselves, but also how they are categorised and perceived by others, both from their 'inside group' and from their 'outside groups' (Beech 2008). Simultaneously it explores the ways in which physical, symbolic, verbal, textual, and behavioural representations become imbued with meaning and are used to structure identities (Beech and Huxham 2003). The complexity and fluidity of interactions, representations and ontological levels in sense-making processes make them problematic to describe and to analyse.

Identity work, if it operates with genes, intersects with narrative, dialogical, and various other symbolical forms of representation. Its exploration needs to consider and utilise the notion that we as human beings are not entities that are contained within ourselves, but rather loose, open, disordered, and unfinalised forms. When engaging with our genetic information, we are not simply absorbing neutral knowledge about the developmental structures of our bodies. Rather, we engage in a process of negotiation about who we are, how we see ourselves, and how we would like others to see us. Genomic knowledge as a practice participates in re-forming our identity, and its own changeability and fluidity plays a huge part in this. Since the genome is not just a biomedical entity or a carrier of information but is also interwoven with symbolic meaning, social values, practice and cultural perception, its disclosure to us—for example, in a counselling situation following a genetic test—re-makes us in a dialogical form. In dialogue with a genetic counsellor, the meaning of our genome becomes 'de-sequenced' into countless bits of meaning re-writing our relationships (for example, with our siblings if an inheritable genetic defect has been detected), our identity (who am I now?) and our possible futures (do I want to have biological offspring?). Such processes of communicative 'de-sequencing" are simultaneously re-sequencing us.

By studying such processes of individual and social sense-making and incorporation at different stages and on different levels we might capture the picture of a 'lived genome' (Rehmann-Sutter and Mahr 2016), which is not a sequence of genetic instructions but a complex sphere of meanings: What does it mean for an individual? For a family? For a society in which more and more people undergo genetic testing? For medical professionals? For policymakers trying to react to the geneticisation of our society?

Hence, genomic sense-making is no longer restricted to people dealing with severe health conditions such as Parkinson's disease. Genetic tests have become available for an increasing range of health conditions. Previously established aetiologies become challenged by additional explanations through a genomic lens, sometimes with astonishing consequences for those who are confronted with new genetic explanations of *their* condition and therefore are a huge part of *themselves*. The ongoing search for genetic influences on health conditions can have a profound impact on the understanding and management of these conditions. This is, for example, the case for health conditions formerly understood to be psychosomatic. An illustration of this is a recent study in *Nature Genetics* that claims to have identified eight genomic risk loci for the psychiatric origins of anorexia nervosa (Watson et al. 2019). This was not only featured enthusiastically as a breakthrough by national and international newspapers such as the *New York Times*, *The Guardian*, and *Le Temps*, but raised new questions on digital patient forums and social health networks such as *PatientsLikeMe*. A controversy unfolded between those who described *their* anorexia as a disease they had to fight day by day to stay alive, and those who aimed to *de-medicalise* the condition as a legitimate form of *being* within the spectrum of human identity—for example, within the 'pro-ana' movement (Hinshaw 2007). The geneticisation of anorexia nervosa might therefore not only impact the scientific description of the disease but also question the identity of those who suddenly find they might be carriers of anorexia genes.

The sequencing of modern life is not only exhausted in the genetic mode, but correlates with the datafication of many areas of life. Even though this book focuses on the levels of translation of genetic information, it must be said that the invisible world of banking algorithms, COVID-19-related proximity tracing, the endless range of health apps, and digital marketing also require de-sequencing (Mahr 2020a, b). Only if this invisible range is dissected analytically, we may all (re-)gain agency towards our modern information society. For this reason, the de-sequencing programme is not only academic, but ultimately also socio-political.

In this volume, we approach personal, social, cultural, and biopolitical identity work with genes from various angles—including sociological, historical, and philosophical accounts as well as first-hand narratives and artistic expressions. The book is divided into five thematic parts. Each of them addresses a specific question and consists of a central text accompanied by related commentaries from other professional or experiential perspectives that aim to further disentangle the sections' theme. The thematic sections of the book examine processes of de-sequencing genetic knowledge in related but distinct contexts of meaning. Their exploration begins with the scientific object itself and then follows different levels where new interpretations and meanings are added. From philosophy to society, to the family as a unit of inheritance and experience, down to the single individual:

1. **Science and medicine**
   What is important about the processing of sequences in 'personalised' or 'systems medicine' and translational research?
2. **Philosophy of biology**
   How is development understood and narrated by biology and philosophy?
3. **Societal contexts**
   How are genetic sequences socially contextualised in big data research and medicine?
4. **Families**
   On what levels is genetic information incorporated into the meaning-making processes by affected families?
5. **Individual experiences**
   How do genes influence illness experiences?

The first section starts with a chapter titled 'Personalised Medicine: Problems of Translation into the Human Domain' by the philosophers of science, Sara Green (University of Copenhagen, Denmark) and Henrik Vogt (postdoctoral fellow at the Institute for Health and Society, Oslo, Sweden). By merging their practical and reflexive experience, they explore how the omnipresent vision of a participatory, preventive, predictive, and personalised medicine, promoted in the scientific and political sphere, fails to address both the clinical practice and societal needs by adapting a narrow and techno-positivist perspective towards patients, users, customers, and clinical experts. But rather than dismissing system medicine as just another regulatory programme ultimately doomed to fail the needs of practice, Green and Vogt propose to integrate awareness about the complexities of human-level needs into the early stages of clinical translation. They show how important it is to disentangle genomic and other forms of biomedical knowledge accurately and adapt translation efforts to target audiences such as patients, study participants, or policymakers. This analysis helps the programme of de-sequencing genetic knowledge to identify the interfaces between the macrosociological level of the science and the medical system with the micro-level dimension of individual actors such as patients or study participants.

The related commentary by the philosopher of biology, Yael Hashiloni-Dolev (Academic College of Tel Aviv–Yaffo) applied the reflection of personalised health prevention to the specific example of prenatal diagnosis. In her arguments, she refers to future parents as educated moral gamblers whose knowledge is extended by the prenatal examination of their foetus; however, this increase in knowledge does not reduce uncertainties. Conversely, it might lead to over-treatment as the choice to terminate a pregnancy or not could become more difficult as a definite outcome remains unpredictable.

The second part (Philosophy of Biology) begins with a chapter on 'Developmental Narratives' by Christoph Rehmann-Sutter (philosopher and bioethicist at the University of Lübeck, Germany). He explores the possibility of a continuity of organic responsivity and psychocorporeal responsivity. To investigate this, Rehmann-Sutter focuses on the discussion of epigenetic development factors in recent biophilosophy. Since this discussion is characterised by a multitude of positions and levels of interpretation, he proposes a model structured according to ontological dimensions in which different forms of responsiveness can be related systematically to each other. In line with the programme of de-sequencing that is the focus of this book, his chapter contributes to translating the complex interconnections of the systems biology approach to microbiological actants (genome, transcriptome, epigenome, etc.) into a form accessible to social analysis.

The first commentary to Rehmann-Sutter's chapter is written by the philosopher of science, Maria Kronfeldner (Central European University, Budapest, Hungary). She scrutinises the meaning of epigenetics for the processes of embodiment and identity work, taking up the old problematic of nature versus nurture. Kronfeldner concludes that we should remain careful when linking embodiment to epigenetics to avoid reducing responsiveness at a molecular level. The second commentary, by the health sociologist Claudine Burton Jeangros (University of Geneva, Switzerland), underlines the importance of space and time when discussing the different layers of developmental narratives. To exemplify the matter of context and period, she mobilises the Chernobyl nuclear accident and the subsequent relevance of social and political institutions for the collective memory. Adopting a life course perspective, she shows how socio-economic factors can manifest under the skin, thereby embodying health inequalities, while Kronfeldner revisits the nature–nurture debate against the backdrop of the embodiment of knowledge.

The third part (Societal Contexts) begins with a text by Barbara Prainsack (Sociology and Bioethics, University of Vienna, Austria). Under the title 'Data Mining in Systems Medicine and the Project of Solidarity: The Interface of Genomics and Society Revisited', she explores the space between futuristic fictional scenarios such as Andrew Niccol's 'Gattaca' (1998), Dave Eggers' 'The Circle' (2013), and the actual, non-fictional world. Focusing on the way digitisation offers hitherto unknown possibilities to personalise health, Prainsack unpacks some assumptions made regarding ethical, legal, and social aspects of collection large sets of health-related data. She proposes avoiding an expanded individualisation of responsibility and reclaiming governance of body data through a new quest for solidarity. Her argument is a strong standpoint to fight for a democratic approach to personalised health, thus protecting citizen who cannot or who choose not to participate. Prainsack hereby focuses the de-sequencing programme on ethically relevant questions of the intangible character of biomedical information. It shows that a reflected handling of such data is of great relevance both for the genesis of an informed sociality and for a social analysis of 'sequences'.

The science studies scholars, Francesco Panese and Luca Chiapperino (both University of Lausanne, Switzerland) in their commentary complement Prainsack's outline with moral and epistemic aspects of personalised medicine. Mobilising Didier Fassin's concept of moral economies, they remind the reader that whatever reality personalised health might become, its realisation is still in a largely debated field of values and imaginations. The bioethicist Gaia Barazzetti (University of Lausanne, Switzerland) questions in her commentary the glorified role of the patient as an engaged and empowered citizen. To build a collective medical future, she advocates for a public discourse on the value of health as a common good.

The fourth part (Families) is introduced by a chapter entitled 'An Ordering of Letters: My Own Personal Genome', written by the digital media scientist and gender researcher Kate O'Riordan (University of Sussex, United Kingdom). Her contribution centres on the dialectical question of how personal genomics, self-tracking, and genetic testing invoke individual responsibility for healthcare, DIY health promotion, and self-governance for prevention on the one hand, while on the other, undoing subjectivity and subjects. She does this via downscaling and focusing on the impact on her own family and their shifting relations in the grip of next-generation genomics, offering a clearly illustrated case study to reveal how much the vision of acquiring useful genomic knowledge might differ from the actual experience of having one's genome sequenced. The latter challenges an experience-based, sometimes messy, translation of information into the concrete life context of affected families.

Her chapter is reflected in the commentary by the sociologist, Barbara Katz Rothman (New York City University, USA) and the social scientist, Silja Samerski (University of Oldenburg, Germany). Katz Rothman takes up the notion of responsibility and examines how our understanding of disease shapes our ideas of whom we should blame for our having a disease. While a genetic component could excuse the concerned person for having a certain disorder, she underlines that the socially attributed stigma will not just vanish. Samerski revisits the complex question about 'how personal a genome can be', and highlights the powerful message underlying genomic information as representing some hidden truth about one's identity. While the genetic forecast does leave many questions unanswered, it leads to a body image of a risk system that needs to be monitored and managed continuously.

In the final part (Individual Experiences), Dana Mahr (historian and sociologist of science, University of Geneva, Switzerland) explores the space of a 'Lived Genome Phenomenology' by analysing the identity narratives of patients and families who deal with chronic inflammatory bowel diseases such as Crohn's disease and ulcerative colitis. These conditions, which were formerly framed as psychosomatic, have recently been re-shaped as clinical disorders with a genetically determined susceptibility, adding *genetic* factors to the explanation of chronic autoimmune diseases of unknown origin. Using a grounded theory approach and interpretative phenomenological analysis, the chapter shines a light on how patients and their families evaluate the changed biomedical explanation of these common diseases, how they make sense out of this in their individual lives, how these processes reframe their perception of biomedical knowledge production, and how this enhances their ability to participate in decision-making processes about their health. This chapter closes the circle of de-sequencing analysis by focusing on the final address of all translations of genetic knowledge that are examined in this book, the individual experience.

Mahr's approach is reflected in a commentary by the philosopher of science, Stephan Kopsieker (formerly Bielefeld University, Germany), who further explores the value of phenomenology when describing and analysing genetic medicine. He underlines the essential contribution of self-told stories in research when examining identity work.

References

Beech, N. (2008). On the nature of dialogic identity work. *Organization, 15,* 51–74. https://doi.org/10.1177/1350508407084485.

Beech, N., & Huxham, C. (2003). Cycles of identity formation in interorganizational collaborations. *International Studies of Management & Organization, 33*(3), 28–52. JSTOR. www.jstor.org/stable/40397570. Accessed 22 September 2020.

Cadwalladar, C. (2014). *The Family Tree: A Novel.* New York: Plume.

Caydunbar. (2013). Patients in waiting. In *Kirk M. Maxey: Blog and Website.* https://kirkmaxey.com/2013/11/22/patients-in-waiting/. Accessed 26 July 2019.

Gibbon, S., & Novas, C. (2007). *Biosocialities, Genetics and the Social Sciences Making Biologies and Identities.* London: Routledge.

Green, R. C., Berg, J. S., Grody, W. W., Kalia, S. S., Korf, B. R., Martin, C. L. et al. (2013). ACMG recommendations for reporting of incidental findings in clinical exome and genome sequencing. *Genetics in Medicine, 15,* 565–574. https://doi.org/10.1038/gim.2013.73.

Hinshaw, S. P. (2007). *The mark of shame: Stigma of mental illness and an agenda for change.* New York: Oxford University Press.

Konrad, M. (2005). *Narrating the New Predictive Genetics: Ethics, Ethnography, and Science.* Cambridge: Cambridge University Press.

Klitzman, R. (2012). *Am I My Genes? Confronting Fate and Family Secrets in the Age of Genetic Testing.* New York: Oxford University Press.

Lauerman, J. (2012). *My DNA Results Spur Alzheimer's Anxiety at $12,000 Cost.* Bloomberg.com. http://www.bloomberg.com/news/articles/2012-11-06/my-dna-results-spur-alzheimer-s-anxiety-at-12-000-cost. Accessed 26 July 2019.

Lippman, A. (1991). Prenatal genetic testing and screening: Constructing needs and reinforcing inequities. *American Journal of Law & Medicine, 17,* 15–50.

Lippman, A. (1999). Choice as a risk to women's health. *Health, Risk & Society, 1,* 281–291.

Mahr, D. (2020a). *The Equilibrium of Control I: Pandemic Exit Strategies in Historical Perspective.* Viral (Blog). https://wp.unil.ch/viral/the-equilibrium-of-control-ii-will-smartphones-help-us-to-overcome-the-sars-cov-2-pandemic/. Accessed 10 June 2020.

Mahr, D. (2020b). *The Equilibrium of Control II: Will Smartphones Help Us to Overcome the SARS-CoV-2 Pandemic?* Viral (Blog). https://wp.unil.ch/viral/the-equilibrium-of-control-ii-will-smartphones-help-us-to-overcome-the-sars-cov-2-pandemic/. Accessed 10 June 2020.

Mahr, D., & Rehmann-Sutter, C. (2015). Information und Partizipation bei Genomstudien. Kommentar zu den „Eckpunkten für eine Heidelberger Praxis der Ganzgenomsequenzierung" der Projektgruppe EURAT. In M. Langanke, P. Erdmann, J. Robienski, & S. Rudnik-Schönenborn (Eds.), *Zufallsbefunde bei molekulargenetischen Untersuchungen. Medizinische, juristische und ethische Perspektiven* (pp. 119–129). Berlin and Heidelberg: Springer.

Marcus, A. D. (2013). Genetic testing leaves more patients living in Limbo. *The Wall Street Journal*. http://www.wsj.com/articles/SB10001424052702303755504579206000052566432. Accessed 26 July 2019.

Munsterhjelm, M. (2014). *Living Dead in the Pacific: Contested Sovereignty and Racism in Genetic Research on Taiwan Aborigines*. Vancouver: University of British Columbia Press.

Nelkin, D., & Lindee, M. S. (2004 [1994]). *The DNA Mystique: The Gene as a Cultural Icon—Conversations in Medicine and Society*. Ann Arbor: University of Michigan.

Rehmann-Sutter, C., & Mahr, D. (2016). The lived genome. In A. Whitehead & J. Richards (Eds.), *Edinburgh Companion to the Critical Medical Humanities* (pp. 87–103). Edinburgh: Edinburgh University Press.

Scully, J. L. (2008). Disability and genetics in the era of genomic medicine. *Nature Reviews Genetics, 9,* 797–802.

Snow, D. A., & Anderson, L. (1987). Identity work among the homeless: The verbal construction and avowal of personal identities. *American Journal of Sociology, 92*(6), 1336–1371.

Vayena, E., & Tasioulas, J. (2013). Genetic incidental findings: Autonomy regained? *Genetics in Medicine, 15,* 868–870. https://doi.org/10.1038/gim.2013.104.

van der Haak, B. (2013). *DNA Dreams*. https://vimeo.com/58285542. Accessed 26 July 2019.

Watson, H. J., Yilmaz, Z., Thornton, L. M., Hubel, C., Coleman, J. R. I., Gaspar, H. A., et al. (2019). *Genome-Wide Association Study Identifies Eight Risk Loci and Implicates Metabo-Psychiatric Origins for Anorexia Nervosa*. Nature Genetics 2019. https://www.nature.com/articles/s41588-019-0439-2#Bib1. Accessed 26 July 2019.

Wolf, S. M., Sobotka, S. P., & Lawrenz, F. P. (2008). Managing incidental findings in human subjects research: Analysis and recommendations. *The Journal of Law, Medicine & Ethics, 36*, 219–248. https://doi.org/10.1111/j.1748-720X.2008.00266.x.

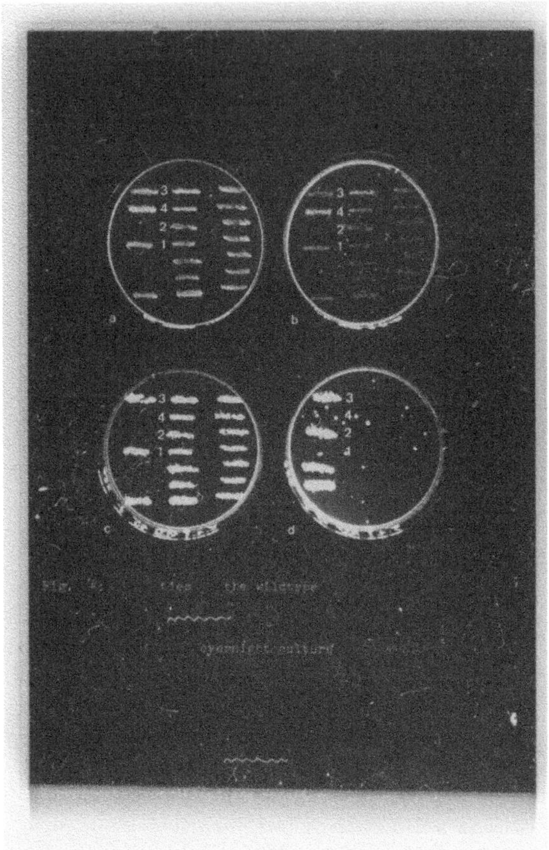

Photo 1 © Grischa Lichtenberger, 0815_11_sc_dia uni b_5_inv_b_ fig 4 ties the wildtype overnight culture

Part I

Science and Medicine

2

Personalised Medicine: Problems of Translation into the Human Domain

Henrik Vogt and Sara Green

2.1 Introduction

We are living in an era of very significant technological advances in describing disease and health. These advances may be spellbinding and lead us to think that huge advances in human health will automatically follow, but the painstaking task of translating the technological advances into clinical relevance is hard. The technology may speak loudly, but that does not mean that the receiver—the patient as a human, living, biological agent—or person—is listening or able to act as expected.

H. Vogt (✉)
Centre for Medical Ethics, Institute of Health and Society, Faculty of Medicine, University of Oslo, Oslo, Norway

Department of Public Health and Nursing, Norwegian University of Science and Technology, Trondheim, Norway

S. Green
Department of Science Education, University of Copenhagen, Copenhagen, Denmark
e-mail: sara.green@ind.ku.dk

D. Mahr and M. von Arx (eds.), *De-Sequencing*, Health, Technology and Society,
https://doi.org/10.1007/978-981-15-7728-4_2

In this chapter we explore translational challenges in the era of personalised medicine (Jain 2015; Shendure et al. 2019). Personalised medicine currently has many synonyms or overlapping concepts. It has also been called precision medicine or P4 medicine (where the Ps stand for 'predictive, preventive, personalised and participatory'), systems medicine, high definition medicine, and deep medicine (Bousquet et al. 2011; Duffy 2015; Torkamani et al. 2017; Topol 2018).

Personalised medicine does for the most part primarily exist as a vision for the future, a construct based on more-or-less credible theoretical assumptions about what biomedicine may achieve (Tutton 2014; De Grandis and Halgunset 2016; Sand 2016; Vogt 2017). One particularly hyped part of this vision is the idea that the new technologies will enable a radically more predictive and effective preventive medicine where individuals can take control of their health. It appears in mainstream media, scientific publications, in political reports, as well as in funding calls in Europe and the USA (Bousquet et al. 2011; Topol 2014; Alyass et al. 2015; Danske Regioner 2015; Centers of Disease Control 2018; PRECeDI 2018). This account of personalised medicine is dominated by an engineering mindset (Hood and Flores 2012). However, as formulated by George Engel, who—as an alternative to a reductionist and mechanistic biomedical model—introduced the biopsychosocial model for medicine in science in 1977, the challenge facing translational biomedicine is '*to be scientific in the human domain*' (Engel 1997).

We examine personalised medicine in light of this challenge, focusing on the challenges of realising personalised, preventive medicine in the real-life settings of clinical practice and society. Our aim is not to reject the potential for medical advances, but to constructively point out that the vision faces great uncertainties and challenges. We argue that the examined preventive strategies risk being translated without a target audience—that is, in ways that do not fit personal and societal contexts—or to be prematurely translated, with the risk of resulting in waste and harm. To improve this situation, we argue that the challenges human-level complexity poses for personalised medicine should be better accounted for, also in initial phases of medical translation.

2.2 The Current Status of Personalised Medicine

The idea of personalised medicine is as old as medicine itself and has previously been tied to *the art of medicine* performed by caring doctors attempting to tailor help according to the characteristics of each patient: biologically and personally, as well as socially. Although his actual opinions as an historical figure are somewhat shrouded in the depths of history, Hippocrates is often quoted as saying that, 'It is more important to know what sort of person has a disease than to know what sort of disease a person has.'[1]

William Osler, a founding father of modern medicine, also noted that, 'if it were not for the great variability among individuals, medicine might as well be a science and not an art.'[2] However, since around 1990 we have witnessed an attempt to make *personalised medicine* a more scientific concept. During the 1990s, in the context of the human genome project, it was appropriated and given a biomedical meaning where *personalisation* becomes the tailoring treatment according to quantitative metrics and biomarkers (Tutton 2014; Jain 2015).

This vision of personalised medicine is often contrasted with present-day evidence-based medicine, presented as a *one-size-fits-all medicine*, where everyone is treated according to a crude population average (Kauffman et al. 2014; Beckman and Lew 2016). However, the main method for personalisation here is *stratification*, which is still a population-based approach. Stratification means that individuals from a large population with similar biological characteristics are aggregated into subgroups and then treated according to what—on average—works for their *stratum* (Beckman and Lew 2016). Additionally, so-called *n-of-1 strategies*, where each person serves as his/her own control, have also been proposed as an underlying method (Schork 2015). In this form, personalised medicine implies a more radical individualisation of medicine in

[1]Hippocrates' quote can be found at BrainyQuotes: https://www.brainyquote.com/quotes/hippocrates_132701, accessed 12 August 2019.

[2]Osler's quote can be found at AZ Quotes https://www.azquotes.com/quote/1396310, accessed 12 August 2019.

the sense that every person is seen to represent a unique case (Boenink 2009).

Since the 1990s, the idea of a biomedical personalised medicine has undergone change following scientific and technological developments (Shendure et al. 2019). One way to track this change is to look at what is meant by the *biological characteristics* according to which people are supposed to receive personalised treatment. At first, the idea was primarily to tailor treatment according to genes (DNA) (Langreth and Waldholz 1999). However, the lesson learnt by biology after the sequencing of the first human genome is that human biology is not genetically determined in any simple way, and is more complex than originally envisioned (Hall 2010). This has led to what we can call a *systems turn* in personalised medicine where its basis shifts from genetics and molecular biology to systems biology (Noble 2006; Hood and Flores 2012). Systems biology realises that the *biological characteristics* according to which one must personalise medicine, are more complex biological networks that include interactions between the genome (DNA) and environmental factors on different levels (Duffy 2016). At this point, personalised biomedicine also appropriated the concept of *holism*, which had previously been a humanistic concept, and filled it with techno-scientific meaning (Vogt et al. 2016a, 2017). The mainstream definition of *precision medicine* (the US government's preferred term) now also reads, 'An emerging approach for disease treatment and prevention that takes into account individual variability in genes, environment, and lifestyle for each person.'[3] In parallel with the scientific and philosophical realisation that one must account for the *whole* of the person in order to personalise, biomedicine has been provided with an array of technologies that produce *big data* and provide a hope that one will actually be able to measure 'the whole of things'. Leroy Hood, a leading biomedical visionary, summarises this idea:

[3]Definition of precision medicine at the US National Library of Medicine: https://ghr.nlm.nih.gov/primer/precisionmedicine/definition, accessed 20 September 2020.

> Ironically many people use the term 'genomic medicine' to denote the medicine of the future – yet in principle genomic medicine is one-dimensional in nature – only encompassing nucleic acid information. Systems medicine, by contrast, is holistic and utilizes all types of biological information – DNA, RNA, protein, metabolites, small molecules, interactions, cells, organs, individuals, social networks and external environmental signals – integrating them so as to lead to predictive and actionable models for health and disease. (Hood and Flores 2012)

The vision of this holistic view is motivated by the availability of several new technologies. The suffix '*-omics*' is used to name technologies that allow for large-scale and automated measurements of *the whole* of some bodily aspect, e.g. transcriptomics (for RNA molecules), proteomics (for proteins), and metabolomics (for metabolites). The measurement of all the traits of the human organism or phenotype is called *phenomics* and the measurement of all environmental exposures *exposomics* (Houle et al. 2010; Wild 2012). Computational advances in artificial intelligence (machine learning) and mathematical modelling offer a promise of making sense of 'the whole of things' (Schork 2019). It is hoped that this will enable more predictive models for preventive purposes (Hood and Flores 2012).

2.3 A Vision of Prevention and Participation

Of central importance for us here, personalised medicine promises to shift medical practice much further from a *reactive mode*—focusing on established, symptomatic disease—to life-long *preventive* medicine focusing on detection of risk, early disease, and proactive intervention (Hood and Flores 2012; Torkamani et al. 2017). In practice, this entails a new form of massive, longitudinal *screening*; that is, detection of early disease or risk factors in asymptomatic people in order to prevent or minimise future disease (Vogt et al. 2016a, 2018; Fiala et al. 2019). Personalised medicine is not only expected to enable the more efficient prevention of rare inherited diseases and cancers, but also of the complex, non-communicable diseases or *lifestyle* diseases that dominate contemporary clinical practice (Bousquet et al. 2011).

A central premise for the realisation of this vision of disease prevention is a cultural shift where people will themselves become very involved as prime contributors to their own monitoring and medical control. This is what lies behind the concept of *participatory medicine* (Topol 2015). Here, personalised medicine very openly enters *the human domain.* Flores et al. (2013) elaborate:

> By adding the 'participatory' component, P4 medicine maximizes the effectiveness of systems medicine by expanding its application out from hospitals and clinics into homes, workplaces and eventually schools. With the addition of self-monitoring (activity, weight and calorie intake) and self-assessments in the participatory component, new quantities and forms of data will be aggregated and mined to generate new insight into health and disease. These insights will drive the development of new technologies, analytic tools and forms of care.

The American National Institutes of Health (NIH) has initiated a research programme called *All of Us*, following this philosophy, where 1 million volunteers are recruited for extensive longitudinal monitoring (National Institute of Health 2018). Only a few studies that reflect the vision described above and supply empirical evidence have been published so far. Our central case example in the present analysis will be the cutting-edge Pioneer 100 Wellness project (P100), featured at the front page of the journal *Nature Biotechnology* and led by personalised medicine visionary Leroy Hood and his collaborator Nathan Price (Price et al. 2017). Another central figure, Craig Venter, and his team subsequently published another study that they called *precision medicine screening* in the *Proceedings of the National Academy of Sciences* (Perkins et al. 2018). Michael Snyder and his team at Stanford University published another study on a 'a longitudinal big data approach for precision health' in the journal, *Nature Medicine* (Rose et al. 2019). Snyder had previously published a study on a single individual (himself) pioneering the massive monitoring of individuals for preventive purposes (Chen et al. 2012). Given the existence of pilot projects which may offer a window to future advances in medicine, we here explore potential translational challenges.

2.4 What Is Translation?

Medical translation refers to the process of developing effective and relevant interventions in both clinic and society from new technologies and basic research. In this analysis, we shall draw on a continuum of translational phases, previously developed by Khoury and colleagues (2007) for public health to genomics, as an analytical tool. This continuum involves four phases:

- Phase 1 translation (T1) involves the move from a basic discovery into a candidate for health application (e.g., medical test/intervention).
- Phase 2 translation (T2) is tied to the evaluation of an application for health practice leading to the development of evidence-based guidelines.
- Phase 3 translation (T3) involves the step from evidence-based guidelines into health practice, through delivery, dissemination, and diffusion.
- Phase 4 translation (T4) is tied to the *real world* health outcomes of a systems medicine application in practice.

The four phases of translation are not sharply delineated in practice. This is especially true of the new big data-based personalised medicine, which further blurs the boundaries between basic and clinical research.[4] As an example, the P100 project is relevant for all four phases of translation:

1. The first aim of P100 involved (among other things) the establishment of methods for analysing multiple sources of health data

[4]All of the three pioneering studies mentioned above are at once basic science studies, observational studies, and clinical studies, where people receive medical interventions aiming simultaneously at basic biomedical discovery, development of biomarkers, therapeutic evaluation, and overall validation of the personalised medicine scheme (Price et al. 2017; Perkins et al. 2018; Rose et al. 2019).

obtained over time from participants in combination with genomic data. This corresponds to phase 1 work.
2. The second aim was to use analytic tools for integrating the diverse data sets and deriving actionable information from their interrelationships. Depending on what is meant by 'actionable information', this corresponds to phases 1, 2, 3, and 4.
3. The third aim was to identify patterns (biomarkers) that correlate with wellness, or transitions between wellness and disease. This corresponds to phases 1 and 2.
4. looseline-1The fourth aim was to study how best to work with and present longitudinal health information to individuals. This corresponds to phase 3.

An important aim of integrating basic and clinical science in the same study is to speed up the process of translation. Implementation which proceeds too fast can, however, result in premature translation (Khoury et al. 2012), and harmful disruption of existing health care systems (Green et al. 2019). After these initial considerations, we shall now proceed with our main argument.

T1: from discovery to candidate health applications
The first phase of translation starts after some basic discovery and involves the move towards the development of a medical test or a treatment that may be tried in the further phases. In the context of preventive personalised medicine, a primary aim is to develop tests to calculate risk and predict the development of disease. This task involves finding the biological characteristics—or biomarkers—according to which medicine may be personalised and will be the most predictive. This is no small task.

As we have outlined above, the biomarkers of personalised medicine have previously been one or a few genetic variants associated with a given disease. Algorithms that mine data from bigger populations have recently allowed for polygenic risk scores that account not only for the existence of a few gene variants but for many as well as their combination (Khera et al. 2018). Other studies focus on screening for rare genetic variants, providing hope that current problems of predicting future disease based on biomarkers may be overcome (Wainschtein et al. 2019).

"Holism" in the context of personalised medicine means that predictions are based on all measurable components of the biocomplexity of disease development (Bousquet et al. 2011). But one cannot simply assume that one can measure and integrate all significant factors that influence human health. This is where translation in preventive personalised medicine collides with the human domain, because the complexity of pathogenesis cannot be separated easily from the complexity of the whole human organism or person trying to adapt in a very complex physical and social environment (Vogt et al. 2014). In his book, *The Mirage of Health*, published in 1959, biologist René Dubos argued that, 'exact science cannot encompass all the human factors involved in health and in disease' (p. 219).

Interestingly, big data science and systems medical approaches have been proposed as a novel form of humanism, able to account for the individual in new ways and perhaps able finally to get past the 'mirage' where science fumbles in grasping the fullness of humanity. Medical scientist Enzo Grossi, writing about artificial intelligence, stated, for example, that 'This approach allows one to partially escape from the probability theory trap in situations where is fundamental to express a judgment based on a single case and favor a novel humanism directed to the management of the patient as an individual subject person' (Grossi 2007: 267). A view that computers can replace doctors, and that 'organisms are algorithms', to be solved by algorithms, also permeates the popular science book *Homo Deus* by historian Yuval Noah Harari. Writing about how supercomputers such as IBM's Watson will know all about his genome, medical history, behaviour and environment, and work tirelessly to process this information in light of all medical literature, he surmises about the future, that 'if you enter medical school today in the expectation of still being a family doctor in twenty years, maybe you should think again. With such a Watson, there is not much need for Sherlocks' (Harari 2017: 315–316).

What Harari presents is, however, a medical dreamscape where he has uncritically accepted the main underlying premise that a machine metaphor of life ('organism is algorithm') is adequate for understanding human beings, and that clinical practice can rest on such understanding alone. However, as Vogt and colleagues write in an article comparing biomedical and humanistic ideas of holism:

> No matter how detailed human phenomes may become, it is still difficult to see how systems biologists can mathematically represent human agency, intentionality and values which are linked to the defining human features of symbolic interaction and narrative. Considering that it is unclear how systems biologists are to quantify a human brain or a living human organism with symbolic behaviour and other social relations, an account of all components of disease complexity, seems outside the scope of quantitative systems medicine. (Vogt et al. 2014)

As noted also by Strand and colleagues, it is not at all a given that elaborate computer models can handle the human complexity that defines clinical practice:

> Medical practice ... involves encounters between human beings. Humans have a richness of mental capacities, such as self-awareness, personality development, creativity, pretending and self-deception, giving them the ability to make self-fulfilling and self-destructive prophecies. Human beings often consciously follow rules, but they can also choose to disobey or change them, and invent new ones. (Strand et al. 2004)

While advocates of personalised medicine may promise new quantitative and objective *metrics* of human wellness and disease, there is nothing to suggest that these metrics will not in the end be influenced by human normative and cultural considerations as to what exactly is an unacceptable condition and what may be considered *health* (Vogt et al. 2016b).

This does not mean that significant discoveries cannot be made, but it does mean that biomedicine is facing an enormous increase in biological complexity as it moves from genes to gene–environment interactions in the discovery and development of health applications. It carries the problem of accounting for this complexity with it to phases 2, 3 and 4.

T2 from health application to evidence-based guidelines
The second phase of translation takes a candidate medical application (test, preventive drug, or treatment) to the point where its use is justified by evidence-based guidelines. We offer some examples below. The T2 phase will overlap with T3 and T4, especially in the new setting of big data screening, which follows people in their real lives, but in theory it takes place in the ideal scenarios of controlled clinical trials.

2.4.1 Evidence-Based Medicine in Flux

The question of how one should develop evidence-based guidelines in personalised medicine is complicated by the fact that it is not entirely clear what evidence-based medicine (EBM) is supposed to mean in this new framework. This is closely tied to the attempt of biomedicine to enter the human domain and account for the complexity of individual, real-world patients and people.

Personalised medicine is often proposed as an alternative to current EBM, which is perceived as being unable to capture sufficiently human complexity and individuality, given the emphasis on randomised controlled trials (RCTs). To quote Kauffman et al. (2014):

> Personalised medicine strives to exploit precisely the information that is randomized away in RCTs by studying one thousand patients or more. Instead of the false certainty obtained by neutralizing patient-specific variation in studies that include as many people as possible, personalised medicine seeks, ideally, a full molecular profiling of an individual patient. The technology is here, and the knowledge of how the human system functions is improving rapidly, such that the collected data cloud around each individual patient can serve as the basis of informed decisions ... The pre-requisites for this approach are high-dimensional measurements that leave no or little unaccounted-for factors that would have to be randomized, and mechanistic knowledge that allows predictive models based on the high-dimensional data obtained for each individual.

In practice, however, we are very far from a situation where analysis leaves 'no or little unaccounted-for factors', and it is questionable as to whether we shall ever get there. At the same time, it is safe to say that the ideal nature of clinical trials is under debate, and that personalised medicine raises questions regarding definitions of ideal or acceptable evidence. EBM is in flux. Stewart and Khoury (2014) summarise the situation:

> New ways of adapting or applying evidentiary standards will be needed, and consideration of the relative weight given to different types of evidence derived from comparative RCTs, observational studies, natural experiments, adaptive trials, pragmatic trials, evidence synthesis, and modeling.

This also means that the idea of what it takes to translate from identified health applications to evidence-based guidelines is changing and not yet quite clear.

It should not be assumed that P4 medicine will easily overcome persistent tensions in medicine, such as between the needs to standardise (to get statistical evidence) and to account for variation (to provide individualised care). Rather, it may be more productive to identify and balance the trade-offs presented by the suggested strategies (Green et al. 2019). In the following, we further unpack complexities that make prediction and evaluation of medical applications particularly challenging in the context of preventive medicine.

2.4.2 Problems of Prediction

Phase T2 involves different forms of research, including observational studies and clinical studies. For medical tests, it seeks to establish their clinical validity and utility. Clinical validity is the accuracy by which a test or algorithm identifies a patient's clinical status and the clinical utility is the benefits and harms that result, from the use of the test (Burke 2014). This brings us to a pivotal set of problems in translation, which we may summarise as *problems of prediction*.

One important question when establishing the predictive values of a test is the prevalence of the disorder. Preventive medicine has been most successful where there is a simple and strong relationship between a biomarker and disease or risk and where there is a high prevalence of disease, which means a high pre-test likelihood of disease (Hamilton et al. 1964; 4S Study Group 1994; Welch et al. 2011, ch. 1). However, in P4 medicine one is proposing to test asymptomatic individuals frequently. Such a population will have a low risk of disease to begin with (so-called pre-test likelihood of disease). This means that the test will have to add a certain amount of predictive power in order to usefully foresee who will actually get the disease (post-test likelihood of disease).

Another important question is, 'What is the natural history of disorders?' To be able to predict the risk implied by some abnormality, and how much it may be altered through an intervention, one has to know how conditions would evolve without such an intervention. Perhaps surprisingly, these facts are not known for many abnormalities. For example, while so-called BRCA-markers are predictive of breast and ovarian cancer, the predictability of these markers can vary among individuals, such as between individuals with and without a family history of cancer, and between different populations (Kuchenbaecker et al. 2017). Thus there are uncertainties associated with the application of tests or markers outside the spectrum of populations studied to validate these applications (e.g. high-risk individuals). This problem in medicine is called *spectrum bias*.

A related translational problem for personalised medicine is that its technologies will be able to detect bodily phenomena that have not been detected previously, and of which we do not yet know the meaning or predictive value. Only long-term studies may establish such meaning. As we mentioned above, this means that one has to tackle not only genetic, but also behavioural and environmental modifiers of disease in human life. To do this, one has to use models that are becoming increasingly complex, and this has important implications for the predictive value of tests. One central problem is the general principle that increasing model complexity with the inclusion of more variables eventually leads to less precision in measurement. As advocates of big data medicine point out themselves, 'it is difficult to reliably estimate many parameters, and correctly infer associations from multiple hypotheses tested simultaneously. As a result, the analysis of both single and integrative omics data is prone to high rates of false-positives due to chance alone' (Alyass et al. 2015). This problem is known as 'the curse of dimensionality', and as 'the bias-variance trade-off' (Fortmann-Roe 2012).

What this means is that precision medicine, which is trying to capture the fullness of human complexity, is inherently and paradoxically challenged by imprecision. Even more fundamentally, complex systems are inherently unpredictable and stochastic, limiting the degree to which future disease can be predicted at all at the level of individual patients (Smith 2011).

2.4.3 Clinical Utility: Problems of Waste and Harm

To evaluate the clinical utility of an application, both positive and negative aspects for patient care must be evaluated. This overlaps with the T3 phase below, but we shall describe some problems here. In general, the waste and harm involved in medical testing and treatment may be both indirect in terms of opportunity costs, where personnel and resources are diverted away from other issues, and direct, in terms of unnecessary side-effects, and costs of diagnostics and treatments such as anxiety and false reassurance. We here highlight three important problems of medical testing of particular relevance when introducing new screening technologies: namely, findings of unknown significance, false positive tests, and overdiagnosis.

2.4.4 Findings of Unknown Significance

Every clinician knows the feeling of having performed a test and not knowing how to interpret it or what to do with a result as it hovers somewhere in the dim landscape between 'clearly significant' and 'clearly insignificant'. Such tests engender more tests to handle the new uncertainty, and thus more potential for waste and harm. In genetics, the finding of variants of unknown significance (VUS), 'routinely failing to reveal useful insights about disease in general or a person's health in particular' has become common (Cooper 2015: 1423). It is likely that this problem will increase with big data testing, both because it may test more variables at one time, and because it may discover new conditions in the human body of which we do not know the significance.

2.4.5 False Positives

False positives are false alarms where a positive test does not actually reflect a pathological reality. As noted, the use of many variables, and many measurements, in diagnostic and predictive algorithms increases the risk of false positive tests. The researchers behind the above-mentioned P100 study also acknowledge that 'It is inevitable that screening thousands of data points will generate false positives, as well as false negatives' (Hood et al. 2015: 2).

2.4.6 Overdiagnosis

However, the most serious problem for the kind of massive screening proposed in P4 medicine may be overdiagnosis (Vogt et al. 2019). This is the detection or diagnosis of abnormalities in asymptomatic people that are not destined to become significant, symptomatic health problems for the individual, either because they disappear spontaneously or the person dies from other causes before such time (Welch et al. 2011). Meta-analyses show that intensified screening is strongly associated with more overdiagnosis and overtreatment (Moynihan et al. 2012). An extreme focus on the early detection of and intervention towards disease in asymptomatic people, as well as widening definitions of what is *actionable*, increases the probability of overdiagnosis (Moynihan et al. 2012).

The basic reasons why overdiagnosis is set to become such a great problem in P4 medicine is the employment of many fragmented variables and measurements in medical testing, each with a low predictive value, as well as the emphasis on continuous surveillance of health states (e.g. via mobile apps). The high definition measurement technologies will detect a host of bodily phenomena of which we do not really know the significance, as we have not previously observed their natural history or variation within the population as a whole. One will be faced with a situation where *everything* abnormal is being detected in the body, but only a very limited amount of this will actually lead to disease, with the other findings representing overdiagnosis.

As an early example, the P100 project mainly contributed the wholesale measurement of a host of markers and their correlations (Price et al. 2017; Vogt et al. 2018). Predictably, this resulted in the diagnosis of multiple risk factors (or *actionable possibilities* as the researchers call them) in all—100%—of 108 previously well (asymptomatic) participants. The researchers involved in the HPWP reported rather cheerfully that 'Many individuals who report that they feel reasonably *well*, may, in fact, have multiple abnormalities in biochemical markers reflecting organ and system dysfunction, nutritional status or other health risk' (Hood et al. 2015: 2). However, labelling everyone as being at risk does not represent a very precise medicine and probably reflects massive overdiagnosis. A similarly imprecise medicine was demonstrated in an early 'precision medicine screening' study conducted by biotechnologist Craig Venter and colleagues (Perkins et al. 2018), and a 'precision health' study by Michael Snyder and colleagues reported in *Nature Medicine* (Rose et al. 2019).

An important translational task is to register not only benefits, but also harm in clinical trials. In the P100 project, critics have raised concerns about the risk of overdiagnosis and overmedicalisation (Vogt et al. 2016a, 2018; Fiala et al. 2019), but the concept of overdiagnosis has yet to enter the preventive, personalised medicine vocabulary, indicating that the field is not properly aware of the translational challenges it faces in the human domain.

2.4.7 Vested Interests

In the human real-world setting of the clinic, an important driver of waste and harm in medicine is vested commercial interests (Moynihan et al. 2012). The billion-dollar start-up company Theranos, which offered a multitude of tests based on analysis of small blood samples, is perhaps the most disturbing and highly publicised example of translation gone astray in the current technological era. While this company for some years until 2015 seemed like the shimmering future and began to offer tests commercially, little or no relevant evidence was published to document the utility of the tests. In 2015, the company's testing-scheme was exposed as unreliable (Ioannidis 2016). This case is extreme. However, at the same time, other commercial enterprises are entering the *theranostics* market that Theranos left open. Arivale, a start-up springing from Leroy Hood's the Institute for Systems Biology and the P100 study, began to offer medical services where health coaches gave advice based on an approach that had not been properly tested. The company then failed (Bishop and Soper 2019).

This type of testing is especially problematic when companies offering tests and counselling profit financially from recommending preventive action without documentation of benefits.

These examples highlight the difficulties of evaluating the clinical utility of new applications. The conclusion that Khoury et al. (2012) reached, that 'lack of information on the clinical utility for most proposed P4 applications produces an evidence dilemma and a conundrum for implementation into practice' is still very relevant as personalised medicine progresses.

T3 from evidence-based guidelines to health practice

Phase 3 in translation starts where one has established that the application is useful under the ideal circumstances of the first clinical trials. With no clear boundary to T2, it is about also achieving clinical utility in the *real world* (Khoury et al. 2007). Here, we shall focus on a critical step, namely the challenge of making people change their behaviour/lifestyle in response to risk assessments in the human domain that is their very complex lives. It is not enough for a test to predict something correctly for it to be useful; it must also be connected to some action that is likely change a prognosis in real-world settings.

First, let us take a historical step back. Individualised preventive medicine has historically certainly had its merits, and there are studies documenting benefits to reap from attempting to change people's lifestyles (Lindström et al. 2003; Li et al. 2008). However, on the whole, decades of research suggest that the prospect of making people change their behaviour is not very credible (Ashenden et al. 1997; Jørgensen et al. 2014; Krogsbøll et al. 2019; Bayer and Galea 2015). While provision of information and lifestyle intervention can be effective in people motivated to improve a health problem, e.g., people with a chronic disease, it typically works poorly on the level of an asymptomatic population. Moreover, considering problems and potential actions in isolation from the lived lives risks overlooking important social factors that make individualised disease prevention difficult. According to Danish epidemiologist Torben Jørgensen and colleagues,

> The reason behind the disappointing results is that health behavior cannot be regarded as isolated. Negative health behavior is dependent on the individual's social context and life conditions. One can therefore not expect that the citizen changes health behavior in a positive direction if the structure of society does not support such behavior. (Jørgensen et al. 2008, our translation)

Proponents of preventive personalised medicine hope that receiving individualised genetic risk information will provide a stronger impetus for behavioural change. But this seems to rest on a strong assumption that risk information consists of value-free facts that are directly translatable into risk-reducing actions. This view not only assumes that it is possible for people to take action, but also that people's goals and values are aligned with the goals of medicine. This expectation is also guiding policymaking in contexts of public health systems. An example is a strategy paper by Danish health authorities, who envision individualised prevention plans based on genome sequencing (Danske Regioner 2015): 'If you can provide early advice to those with highest risk, you might may be able to better motivate them to a change of life-style' (Danske Regioner 2015: 6, our translation). However, these expectations go directly against existing evidence. A systematic review of existing evidence and meta-analysis concluded:

> Expectations that communicating DNA based risk estimates changes behaviour is not supported by existing evidence. These results do not support use of genetic testing or the search for risk-conferring gene variants for common complex diseases on the basis that they motivate risk-reducing behaviour. (Hollands et al. 2016)

Nevertheless, hopes for preventive medicine based on systems medicine and quantified-self technologies have not been deterred by these setbacks. It is hoped that feedback from a more constant and regular monitoring of systems medicine will engender change. For example, Hood and colleagues behind the P100 study argue that: 'The ongoing interaction between participants and health coach and each other through social networking and community creates relationship-based accountability—increasing motivation for long-term lifestyle change' (Hood et al. 2015). But the results from the P100 study seem not to offer support for this. While the pioneering P4 medicine study was touted as showing that 'behavioural coaching informed by personal data helped participants to improve clinical biomarkers', there was no control group to establish this fact (for more elaborate criticisms, see Vogt et al. 2018; and Fiala et al. 2019). It is also interesting to note that only 64 per cent of the participants managed to use a Fitbit monitoring device to measure activity levels for a minimum of 40 days during a nine-month period. Compliance with sleep tracking was even lower (Price et al. 2017). This highlights that, even in a population of highly motivated and socio-economically relatively privileged individuals, the issue of compliance remains a hurdle. The translation of basic science into individualised behavioural change may thus be a translation without an audience. What the future brings remains to be seen. However, other pioneering studies also suggest that the picture will not change.

In 2016, a randomised controlled trial was performed by personalised medicine visionary Eric Topol and colleagues at Scripps Research in the USA (previously known as The Scripps Research Institute). It tested an extensive self-monitoring system and follow-up over six months in participants with common, chronic health conditions. The result showed no short-term effects on health care utilisation or costs (Bloss et al. 2016). The year after, Jakicic and colleagues published a two-year study of 471 overweight/obese individuals (Jakicic et al. 2017). All participants received a low-calorie diet, and were prescribed increased physical activity and group counselling sessions. After six months, the participants were randomised into two groups, where a standard intervention group initiated self-monitoring of diet and physical activity using a website; and an enhanced intervention group were provided with a wearable device and web interface to monitor diet and physical activity. Interestingly, the standard intervention group achieved twice the weight loss as those provided with a P4-medicine-style intervention. The reasons for this are not entirely clear, but it highlights the complexities of achieving preventive control in people's real lives.

2.4.8 Social Level Damage

Aside from concerns that the intervention may lead to waste and harm at the individual level (for people participating in preventive programmes), the lack of established utility may also have harmful effects at the societal level. A serious and very likely downside is indirect harm stemming from opportunity cost. The hype around personalised medicine may distract attention and draw resources away from more effective public health actions. Given limitations on health care resources, it is a considerable concern to ensure that resources are not wasted on (over)medicalisation of healthy people but prioritised for those in need of healthcare. As Burke and Trinidad (2011: 1) stress, 'P4 medicine cannot solve the root problem: the need for political and public health action to improve the life chances of disadvantaged people. In this context, a realistic assessment of the prospects for systems biology is sorely needed.'

In the wave of enthusiasm over new health technologies it is often overlooked that many healthcare problems are likely to be better solved by addressing the environmental and social challenges facing individuals. Back in 2008, public health researchers pointed to a 'paradigm shift in health promotion and disease prevention' highlighting that "the primary strategy should be structural" (Jørgensen et al. 2008). That is, rather than ineffectively, and perhaps sometimes unethically, confronting individual citizens with (often uncertain) numbers about individual disease risk and recommended preventive actions, effects are more likely to result from changes in structural and societal constraints for citizens at the population level. These lessons from public health and epidemiology are unfortunately often ignored.

Ironically, preventive personalised medicine has even been promoted as a particularly potent solution to global health challenges in the developing world (Bousquet et al. 2011). The idea here is that, in the same way as developing countries may jump straight from no communication technology (traditional telephones) to advanced smart-phones, they may bypass a traditional health care system and move straight to a digital doctor in the palm of their hand. However, it is unfounded to assume that what the developing world needs is advanced mobile health when the most pressing needs are basics such as clean water and adequate food, fundamental primary care, a raising of basic living standards, and freedom from oppression, violence, and other major stressors.

We mention one example that illustrates the joint concerns of overdiagnosis and opportunity cost. Various teams of researchers have in recent years proposed smart-phone cameras and apps as an easy way to self-diagnose jaundice in babies (yellow skin from a waste-product called bilirubin) which, if untreated, may lead to death or brain damage (Furman 2017). Assuming that the test is accurate, the benefits may look obvious. However, only a small proportion (1/1000) of the approximately 120,000,000 children born with yellow skin from high bilirubin levels develop associated health problems. Testing all babies south of the Sahara, where the problem is most common, would likely lead to massive overdiagnosis and place a significant burden on a healthcare system that currently does not even deliver basic health care. In this context we cannot assume that babies who would benefit from diagnosis and treatment would have access to healthcare at all, or that caregivers would be able to turn the findings into useful action. What is the use of an app if one cannot get help, or if the socio-economic circumstances are such that some children are neglected? While this may be an extreme example, it illustrates the need to account for societal and social contexts when estimating the likely effect of medical applications.

2.4.9 Social and Cultural Iatrogenesis

Arguably, participatory personalised medicine and self-monitoring technologies most evidently expand the ability of people to self-medicalise, resulting in 'participatory medicalisation', where consumers are the main drivers of the medicalisation process (Vogt et al. 2016a). As highlighted by Vogt et al. (2016a), a looming danger here is what Ivan Illich termed social and cultural iatrogenesis (Illich 1975): it may displace other collectively learned ways of dealing with disease and health. Such an all-encompassing medical scheme risks incorporating aspects of human life into medicine that do not belong there, but rather on a social or political level.

The *womb-to-tomb* vision of preventive personalised medicine has the potential to redefine our very personal life-stories as techno-scientific *bio-narratives*. If we accept the philosopher Hans-Georg Gadamer's definition of health as the ability to forget that one is healthy (Gadamer 1996: vii), health would seem hard to obtain in a society following the ideals of preventive personalised medicine (Vogt et al. 2016a).

T4: conclusion—from practice to population health impact
The final phase is where the three previous phases sum up to yield a significant, positive impact on population health as measured by outcomes research. We shall use this phase to conclude our discussion.

The promise of personalised medicine is very significant gains in the utility of disease prevention, and results for the health of the population. At the same time, the massive monitoring and attempts at continual control throughout life that is proposed in systems medicine represents an unprecedented degree of medicalisation with inherent risks of waste and harm (e.g. from overdiagnosis as described above). For the balance of benefits versus waste and harm to become positive, this radical expansion seems to presuppose very significant gains in the overall utility of preventive medicine (Green and Vogt 2016; Vogt 2017). However, there are reasons to believe this promise is unreliable (Bayer and Galea 2015; Vogt 2017).

First, remember that the population in Western countries in general is quite healthy and with a high life expectancy. The main possible exception is the USA, which has had a negative development in recent years, mainly for political and socio-economic reasons outside the scope of personalised medicine (Bernstein 2018). In general, the prospect of improving public health in a population that is already quite healthy is hard and necessarily has a limited upside.

Personalised medicine is introduced at a time when there are serious concerns as to the overall utility (balance of benefits and harms) of 'too much medicine', especially with regard to preventive medicine's expanding efforts.[5] Without further documentation, Hood et al. promise that the benefits of their approach will 'far outweigh any possible harms' (Hood et al. 2015: 3). However, this has so far not been shown to be likely, and preventive personalised medicine thus risks premature translation. Precision medicine therefore risks continuing what David Sackett, evidence-based medicine pioneer, called 'the arrogance of preventive medicine':

> Preventive medicine displays all 3 elements of arrogance. First, it is aggressively assertive, pursuing symptomless individuals and telling them what they must do to remain healthy ... Second, preventive medicine is presumptuous, confident that the interventions it espouses will, on average, do more good than harm to those who accept and adhere to them. Finally, preventive medicine is overbearing, attacking those who question the value of its recommendations. (Sackett 2002: 363)

Personalised medicine is associated with a range of drivers behind waste and harm in medicine, such as new technology that allows for increasingly sensitive measurements of bodily factors, a widened scope of medicine through an increasing focus in healthy people, and health consumerism (Welch et al. 2011; Moynihan et al. 2012; Heath 2013). While companies are jumping the gun in offering personalised medicine services, evidence is lacking from long-term studies. Personalised medicine is thus starting to create a disorderly clinical situation where no one can meaningfully translate all the new, big data.

We have pointed out how the translation of basic science into individualised behavioural change may be translation without an audience. Since the main translational hurdle is to change the situation for those most in need of health care, a more constructive approach would be to focus attention on to translational strategies that do have an audience. As Khoury et al. (2012) emphasised, the social complexity and challenges

[5]See, for example, *The BMJ* (*British Medical Journal*), 'Too much medicine', at http://www.bmj.com/too-much-medicine.

encountered in the clinic and in society should be accounted for at the outset, and not acknowledged only at the final phase, where translation may be blocked and resources wasted.

Acknowledgments Funding: Green's research is supported by The Carlsberg Foundation (Semper Ardens grant CF17-0016) and the Danish Research Council for Independent Research (Project 1-grant 0132-00026B).

References

4S Study Group. (1994). Randomised trial of cholesterol lowering in 4444 patients with coronary heart disease: The Scandinavian Simvastatin Survival Study (4S). *Lancet, 344*(8934), 1383–1389.

Alyass, A., Turcotte, M., & Meyre, D. (2015). From big data analysis to personalised medicine for all: Challenges and opportunities. *BMC Medical Genomics, 8,* 33.

Ashenden, R., Silagy, C., & Weller, D. (1997). A systematic review of the effectiveness of promoting lifestyle change in general practice. *Family Practice, 14*(2), 160–176.

Bayer, R., & Galea, S. (2015, August 6). Public health in the precision-medicine era. *The New England Journal of Medicine, 373*(6):499–501. https://scite.ai/reports/10.1056/nejmp1506241.

Beckmann, S., & Lew, D. (2016). Reconciling evidence-based medicine and precision medicine in the era of big data: Challenges and opportunities. *Genome Medicine, 8,* 134. https://doi.org/10.1186/s13073-016-0388-7.

Bernstein, L. (2018, November 29). U.S. life expectancy declines again, a dismal trend not seen since World War I. *Washington Post*. Accessed 3 July 2019.

Bishop, T., & Soper, T. (2019, April 24). *Scientific wellness startup Arivale closes abruptly in 'tragic' end to vision to transform personal health.* Geekwire. https://www.geekwire.com/2019/scientific-wellness-startup-arivale-closes-abruptly-tragic-end-vision-transform-personal-health. Accessed 3 July 2019.

Bloss, C. S., Wineinger, N. E., Peters, M., Boeldt, D. L., Ariniello, L., Kim, J. Y., et al. (2016). A prospective randomized trial examining health care utilization in individuals using multiple smartphone-enabled biosensors. *PeerJ, 4,* e1554.

Boenink, M. (2009). Tensions and opportunities in convergence: Shifting concepts of disease in emerging molecular medicine. *Nanoethics, 3*(3), 243–255.

Bousquet, J., Anto, J. M., Sterk, P. J., Adcock, I. M., Chung, K. F., Roca, J. et al. (2011, July 6). Systems medicine and integrated care to combat chronic noncommunicable diseases. *Genome Medicine, 3*(7), 43. https://doi.org/10.1186/gm259.

Burke, W. (2014, April 24). Genetic Tests: Clinical Validity and Clinical Utility. *Current Protocols in Human Genetics, 81,* 9.15.1–9.15.8. https://doi.org/10.1002/0471142905.hg0915s81.

Burke, W., & Trinidad, S. B. (2011). Systems medicine and the public's health. *Genome Medicine, 3*(7), 47.

Centers of Disease Control. (2018). *Precision Medicine: What Does It Mean for Your Health?* https://www.cdc.gov/features/precision-medicine/index.html. Accessed 3 July 2019.

Chen, R., Mias, G. I., Li-Pook-Than, J., Jiang, L., Lam, H. Y., Chen, R. (2012, March 16). Personal Omics Profiling Reveals Dynamic Molecular and Medical Phenotypes. *Cell, 148*(6), 1293–307. https://doi.org/10.1016/j.cell.2012.02.009.

Cooper, G. M. (2015). Parlez-vous VUS? *Genome Research, 25*(10), 1423–1426.

Danske Regioner. (2015). *Personalised Medicine and Individualized Treatment: Roadmap Towards a Unified Danish Strategy.* Copenhagen, Danish Regions—in Danish.

De Grandis, G., & Halgunset, V. (2016). Conceptual and terminological confusion around personalised medicine: A coping strategy. *BMC Medical Ethics, 17,* 43.

Dubos, R. J. (1959). *Mirage of Health, Utopias, Progress, and Biological Change.* New York: Harper.

Duffy, D. J. (2016, May). Problems, challenges and promises: Perspectives on precision medicine. *Brief Bioinform, 17*(3), 494–504. https://doi.org/10.1093/bib/bbv060. Epub 2015 August 5.

Engel, G. L. (1977). The need for a new medical model: A challenge for biomedicine. *Science, 196* (4286), 129–136.

Engel, G. L. (1997). From biomedical to biopsychosocial: Being scientific in the human domain. *Psychosomatics, 38*(6), 521–528.

Fiala, C., Taher, J., Diamandis, E. P. (2019, March 26). Benefits and harms of wellness initiatives. *Clinical Chemistry and Laboratory Medicine*. https://doi.org/10.1515/cclm-2019-0122.

Flores, M., Glusman, G., Brogaard, K., Price, N. D., & Hood, L. (2013). P4 medicine: How systems medicine will transform the healthcare sector and society. *Per Med, 10*(6), 565–576.

Fortmann-Roe, S. (2012). *Understanding the Bias-Variance Tradeoff*. http://scott.fortmann-roe.com/docs/BiasVariance.html.

Furman, L. (2017). Bilirubin phone apps—Our future calls! *Journals blog*. https://www.aappublications.org/news/2017/08/28/Bilirubin-phone-apps-pediatrics-7-28. Accessed 3 July 2019.

Gadamer, H.-G. (1996). *The Enigma of Health*. Cambridge: Polity.

Green, S. M. E., & Vogt, H. (2016). Personalizing medicine: Disease prevention in silico and in socio. *Humana. Mente Journal of Philosophical Studies, 30*, 105–145.

Green, S., Carusi, A., & Hoeyer, K. (2019, May 18). Plastic diagnostics: The remaking of disease and evidence in personalised medicine. *Social Science & Medicine*. S0277-9536(19)30287-4. https://doi.org/10.1016/j.socscimed.2019.05.023.

Grossi, E. (2007). Technology transfer from the science of medicine to the real world: The potential role played by artificial adaptive systems. *Substance Use & Misuse, 42*(2–3), 267–304.

Hall, S. S. (2010). Revolution postponed. *Scientific American, 303*(4), 60–67.

Hamilton, M., Thompson, E. M., & Wisniewski, T. K. (1964). The role of blood-pressure control in preventing complications of hypertension. *Lancet, 1*(7327), 235–238.

Harari, Y. N. (2017). *Homo Deus: A Brief History of Tomorrow* (Kindle ed.). Harper.

Heath, I. (2013). Overdiagnosis: when good intentions meet vested interests—An essay by Iona Heath. *BMJ, 347*, f6361.

Hollands, G. J., French, D. P., Griffin, S. J., Prevost, A. T., Sutton, S., King, S., et al. (2016). The impact of communicating genetic risks of disease on risk-reducing health behaviour: Systematic review with meta-analysis. *BMJ, 352*, i1102.

Hood, L., & Flores, M. (2012, September 15). Flores M. A personal view on systems medicine and the emergence of proactive P4 medicine: Predictive, preventive, personalised and participatory. *New Biotechnology, 29*(6), 613–624. https://doi.org/10.1016/j.nbt.2012.03.004.

Hood, L., Lovejoy, J. C., & Price, N. D. (2015). Integrating big data and actionable health coaching to optimize wellness. *BMC Medicine, 13,* 4.

Houle, D., Govindaraju, D. R., & Omholt, S. (2010). Phenomics: The Next Challenge. *Nature Reviews Genetics, 11*(12), 855–866.

Illich, I. (1975). *Medical Nemesis: The Expropriation of Health.* London: Calder & Boyars.

Ioannidis, J. P. (2016, July 26). Stealth research and Theranos: Reflections and update 1 year later. *JAMA, 316*(4), 389–390. https://doi.org/10.1001/jama.2016.6986.

Jain, K. K. (2015). *Textbook of Personalised Medicine* (2nd ed.). New York, Heidelberg, Dordrecht, and London: Springer.

Jakicic, J. M., Davis, K. K., Rogers, R. J., King, W. C., Marcus, M. D., Helsel, D. (2016, September 20). Effect of wearable technology combined with a lifestyle intervention on long-term weight loss: The IDEA randomized clinical trial. *JAMA, 316*(11), 1161–1171. https://doi.org/10.1001/jama.2016.12858.

Jørgensen, T., Glümer, C., & Pisinger, C. (2008). *A Paradigm Shift in Prevention.* Region Hovedstaden. Nyt om forebyggelse. Article in Danish. https://dokodoc.com/nyt-om-forebyggelse-region-hovedstaden-forskningscenter-for.html.

Jørgensen, T., Jacobsen, R. K., Toft, U., Aadahl, M., Glumer, C., & Pisinger, C. (2014). Effect of screening and lifestyle counselling on incidence of ischaemic heart disease in general population: Inter99 randomised trial. *BMJ, 348,* g3617.

Kauffman, S., Hill, C., Hood, L., & Huang, S. (2014). Transforming medicine: A Manifesto. *Scientific American Worldview,* 24, 28–31.

Khera, A. V., Chaffin, M., Aragam, K. G., Haas, M. E., Roselli, C., Choi, S. H. (2018, September). Genome-wide polygenic scores for common diseases identify individuals with risk equivalent to monogenic mutations. *Nature Genetics, 50*(9), 1219–1224. https://doi.org/10.1038/s41588-018-0183-z.

Khoury, M. J., Gwinn, M., Yoon, P. W., Dowling, N., Moore, C. A., & Bradley, L. (2007, October). The continuum of translation research in genomic medicine: How can we accelerate the appropriate integration of human genome discoveries into health care and disease prevention? *Genetics in Medicine, 9*(10), 665–674.

Khoury, M. J., Gwinn, M., Glasgow, R. E., & Kramer, B. S. (2012). A population perspective on how personalised medicine can improve health. *American Journal of Preventive Medicine, 42*(6), 639.

Krogsbøll, L. T., Jørgensen, K., & Gøtzsche, P. C. (2019). General health checks in adults for reducing morbidity and mortality from disease. *Cochrane Database of Systematic Reviews*, Issue 1. Art. No.: CD009009. https://doi.org/10.1002/14651858.CD009009.pub3.

Kuchenbaecker, K. B., Hopper, J. L., Barnes, D. R., Phillips, K. A., Mooij, T. M., Roos-Blom, M. J. et al. (2017, June 20). Risks of breast, ovarian, and contralateral breast cancer for BRCA1 and BRCA2 mutation carriers. *JAMA, 317*(23), 2402–2416. https://doi.org/10.1001/jama.2017.7112.

Langreth, R., & Waldholz, M. (1999). New era of personalised medicine: Targeting drugs for each unique genetic profile. *Oncologist, 4*(5), 426–427.

Li, G., Zhang, P., Wang, J., Gregg, E. W., Yang, W., Gong, Q. (2008, May 24). The long-term effect of lifestyle interventions to prevent diabetes in the China Da Qing Diabetes Prevention Study: A 20-year follow-up study. *Lancet, 371*(9626), 1783–1789. https://doi.org/10.1016/S0140-6736(08)60766-7.

Lindström, J., Louheranta, A., Mannelin, M., Rastas, M., Salminen, V., Eriksson, J., Uusitupa, M. et al. (2003, December 26). Finnish Diabetes Prevention Study (DPS): Lifestyle intervention and 3-year results on diet and physical activity. *Diabetes Care, 26*(12), 3230–3236.

Moynihan, R., Doust, J., & Henry, D. (2012). Preventing Overdiagnosis: How to Stop Harming the Healthy. *BMJ, 344,* e3502.

National Institutes of Health. (2018, March 28). *All of Us Research Program, National Institutes of Health Core Protocol v1.7.*

Noble, D. (2006). *The Music of Life: Biology Beyond the Genome.* Oxford: Oxford University.

Perkins, B. A., Caskey, C. T., Brar, P., Dec, E., Karow, D. S., Kahn, A. M. et al. (2018). Precision medicine screening using whole-genome sequencing and advanced imaging to identify disease risk in adults. *Proceedings of the National Academy of Sciences of the United States of America, 115*(14), 3686–3691. https://doi.org/10.1073/pnas.1706096114. Epub 2018 March 19.

PRECeDI. (2018). *How to Integrate Personalised Medicine into Prevention? Recommendations from the Personalised Prevention of Chronic Diseases (PRECeDI) Consortium.* https://ephconference.eu/repository/projects/Final_Recommendations_How_to_integrate_Personalized_Medicine_into_Prevention.pdf

Price, N. D., Magis, A. T., Earls, J. C., Glusman, G., Levy, R., Lausted, C. et al. (2017, August). A wellness study of 108 individuals using personal, dense, dynamic data clouds. *Nat Biotechnol, 35*(8):747–756. https://doi.org/10.1038/nbt.3870. Epub 2017 July 17.

Rose, S. M., Contrepois, K., Moneghetti, K. J., Zhou, W., Mishra, T., Mataraso, S. (2019). A longitudinal big data approach for precision health. *Nature Medicine, 25*(5), 792–804. https://doi.org/10.1038/s41591-019-0414-6.

Sackett, D. L. (2002, August 20). The Arrogance of Preventive Medicine. *CMAJ*, *167*(4), 363–364.

Sand, M. (2016). Responsibility and visioneering—Opening Pandora's box. *Nanoethics, 10,* 75–86. https://doi.org/10.1007/s11569-016-0252-7.

Schork, N. J. (2015, April 30). Personalised medicine: Time for one-person trials. *Nature, 520*(7549), 609–611. https://doi.org/10.1038/520609a.

Schork, N. J. (2019). Artificial intelligence and personalised medicine. *Journal of Cancer Treatment and Research, 178,* 265–283. https://doi.org/10.1007/978-3-030-16391-4_11.

Shendure, J., Findlay, G. M., Snyder, M. W. (2019, March). Genomic medicine-progress, pitfalls, and promise. *Cell, 177*(1), 45–57. https://doi.org/10.1016/j.cell.2019.02.003.

Smith, G. D. (2011). Epidemiology, epigenetics and the 'Gloomy Prospect': Embracing randomness in population health research and practice. *International Journal of Epidemiology, 40*(3), 537–562.

Stewart, A., & Khoury, M. J. (2014). *Is Evidence-based Medicine the Enemy of Genomic Medicine?* CDC Blogs. https://blogs.cdc.gov/genomics/2014/02/13/is-evidence-based/. Accessed November 27, 2017.

Strand, R., Rørtveit, G., & Schei, E. (2004). Complex systems and human complexity in medicine. *Complexus, 5*(2), 2–6.

Topol, E. J. (2014, March 27). Individualized medicine from prewomb to tomb. *Cell, 157*(1), 241–253. https://doi.org/10.1016/j.cell.2014.02.012.

Topol, E. J. (2015). *The Patient Will See You Now: The Future of Medicine Is in Your Hands*. Basic Books.

Topol, E. J. (2019). *Deep Medicine: How Artificial Intelligence Can Make Healthcare Human Again.* Basic Books.

Torkamani, A., Andersen, K. G., Steinhubl, S. R., Topol, E. J. (2017, August 27). High-definition medicine. *Cell, 170*(5), 828–843. https://doi.org/10.1016/j.cell.2017.08.007.

Tutton, R. (2014). *Genomics and the Reimagining of Personalised Medicine.* Farham: Ashgate.

Vogt, H., Ulvestad, E., Eriksen, T. E., & Getz, L. (2014). Getting personal: Can systems medicine integrate scientific and humanistic conceptions of the patient? *Journal of Evaluation in Clinical Practice, 20,* 942–952.

Vogt, H., Hofmann, B., & Getz, L. (2016a). The new holism: Systems medicine and the medicalization of health and life itself. *Medicine Health Care and Philosophy, 19*(2), 307–323.

Vogt, H., Hofmann, B., & Getz, L. (2016b). Personalised medicine: Evidence of normativity in its quantitative definition of health. *Theoretical Medicine and Bioethics, 37*, 401–416.

Vogt, H. (2017). *Systems Medicine as a Theoretical Framework for Primary Care Medicine: A Critical Analysis* (PhD thesis). NTNU Open. https://ntnuopen.ntnu.no/ntnu-xmlui/handle/11250/2441891.

Vogt, H., Green, S., & Brodersen, J. (2018, August 6). Precision Medicine in the Clouds. *Nat Biotechnol, 36*(8), 678–680. https://doi.org/10.1038/nbt.4210.

Vogt, H., Green, S., Ekstrøm, C., & Brodersen, J. (2019). Much more medicine: Why big data screening means more overdiagnosis. *BMJ, 366,* 15270.

Wainschtein, P., Jain, D. P., Yengo, L., Zheng, Z., Cupples, L. A, Shadyab, A. H. et al. (2019). *Recovery of Trait Heritability from Whole Genome Sequence Data.* BioRxiv. Preprint. https://doi.org/10.1101/588020.

Welch, H. G., Schwartz, L., & Woloshin, S. (2011). *Overdiagnosed: Making People Sick in the Pursuit of Health.* Boston, MA: Beacon Press.

Wild, C. P. (2012). The exposome: From concept to utility. *International Journal of Epidemiology, 41*(1), 24–32.

3

Contemporary Future Parents: From Tentative Pregnancy and Moral Pioneers to Educated Moral Gamblers

Yael Hashiloni-Dolev

'Lost in Translation: Personalised disease prevention in the era of big data and systems medicine' by Vogt and Green, focuses on the challenges of this new medical era for the general public, not patients, as disease prevention targets asymptomatic people. In my comment, I wish to address some of their concerns regarding the fundamental problems, uncertainties and challenges posed by translating predictive big data into useful healthcare applications in the specific field of prenatal diagnosis. Speaking of diagnosis of pre-born humans, who are obviously not patients, I will not return to the broad questions of enhancement, eugenics, perfection, slippery slope, or designing of future generations, all very well studied and contemplated ethical and social issues in this field.

My commentary will follow another strand of scholarly work, which focused on the experiences of expecting parents, mainly, of course, pregnant women, since the introduction of amniocentesis as a rather

Y. Hashiloni-Dolev (✉)
Ben-Gurion University of the Negev, Be'er-Sheva, Israel
e-mail: yaelhd@bgu.ac.il

D. Mahr and M. von Arx (eds.), *De-Sequencing*, Health, Technology and Society,
https://doi.org/10.1007/978-981-15-7728-4_3

common diagnostic test during pregnancy, towards the end of the 1970s. This old/new effort to predict the future child's health, while being a very private experience, occurs in certain cultural, medical, economic, religious, and technical contexts. All of which have been documented as the background for pregnant women's medical encounters and decision-making processes concerning testing and selective abortions, in the era of the diagnosable embryo/foetus. Katz-Rothman, as early as 1986, and Rapp in 1999, wrote classic works about this, then new, medical experience of pregnancy. In the very early days of amniocentesis, Katz-Rothman suggested that mothers were now expected to see their children as products of conception, which have to be carefully examined for quality control before leaving 'the factory'. This was seen as part of the general process of the commodification of the human body. She further claimed that what were then new genetic reproductive techniques, also posed contradictory demands on women: they were expected to love their child and take care of it from the moment of conception, but also to be willing to abort it if a genetic problem was detected. Hence she coined the term 'tentative pregnancy', to describe this ambivalence (Katz-Rothman 1986).

Thirteen years later, Rayna Rapp (1999) used the term 'moral pioneers' to complement the description of modern pregnant women's emotional and cognitive experience. Situated on a research frontier of the expanding capacity for prenatal genetic diagnosis, Rapp described future mothers as being forced to judge the quality of their own foetuses, and to make concrete and embodied decisions about the standards for entry into the human community (Rapp 1999). Writing more than thirty years after Katz-Rothman, and almost twenty years later then Rapp, can we speak of a change, and if so, of what kind? Contemporary advances in molecular genetics allow detailed investigations of the foetal genome. Chromosomal microarray analysis (CMA), whereby small gains and losses of genetic material are identified, have become the first line of genetic investigation in pregnancies with foetal abnormalities detected on ultrasound. Likewise, a growing number of centres worldwide are offering CMA to all women undergoing invasive prenatal testing. Additionally, whole-exome-testing (WES), whereby the encoding part of the genome is examined, have been introduced in pregnancies with structural anomalies. The main benefit of advanced genomic tests (CMA, WES) is their higher detection rate, which while not desired, is helpful for patients and doctors in managing the pregnancy. Also, more accurate information about the risks of a recurrence can help plan future pregnancies, or take decisions regarding preimplantation genetic diagnosis (PGD).

Nevertheless, the higher resolution in which the genome is examined in advanced genomic tests, also has its downsides. The advanced tests are profoundly different from classical ones, since they are not intended to look for a condition suspected following clinical examinations of the phenotype. Rather, more like throwing a fishing net into the sea, they are scanning the whole 'genomic ocean' and waiting to see what comes up. Therefore it is the genotype and not the phenotype that directs the search. When foetuses are examined, and no clinical check-up is possible, full diagnosis and risk assessment are even more difficult to make. Thus detection of a genetic problem, which in the early days of amniocentesis was mostly about having (or not having) an extra chromosome 21, has become far more complex. In supposedly simple cases, a definite diagnosis can be reached. But even then, in many instances, the severity of the condition varies, and future possible treatment is unpredictable.

As Vogt and Green write, increasing model complexity, does not lead to better prediction. On the contrary, it makes prediction actually more difficult. So it is a common event that probabilistic rather than deterministic information of knowledge is reached. Such information includes variants with uncertain clinical significance (VUS), meaning it is impossible to determine whether they are meaningless or pathogenic. Such findings are obviously stress factors for parents and professionals alike, as what knowledge has brought along is nothing but uncertainty. Moreover, this uncertainty may stigmatise the foetus, and cause anxiety and over-diagnosis during pregnancy and possibly post-birth for years to come.

Other types of findings commonly identified in advanced genomic tests are incidental ones, such as risks for adult-onset conditions (e.g. deletion of a cancer susceptibility gene) and mutations in genes associated with X-linked and autosomal recessive diseases.

Genetic knowledge, like all medical knowledge, can never be completely certain. None the less, current tendencies force future parents to deal increasingly with probabilistic uncertain knowledge, without being able to predict anything about the future. For example, what can parents know about their children's future lifestyles, or about yet to be developed relevant therapies, or the state of the environment?

This query is, of course, especially relevant for late onset conditions. A greater risk for developing one day in the far future, common diseases such as diabetes, heart disease or cancer, is very hard to think through. That is since we are ignorant about our children's future life circumstances; will they live long enough to develop the disease? Will they be among the 50% who do or do not develop it? At what age? What will be the severity of the disease, what other life style and risk factors will be involved, and what will future therapies look like?

And so, in today's world, parents might be faced with such a hypothetic question; what is preferable, or rather, more/less risky? To be born with a 50% chance of suffering juvenile diabetes, or a 70% chance of getting breast cancer around the age of forty? What can a supposedly rational person/future parent make of such an informed choice?

Would there still be any 'perfect' babies to choose among? Or only future children with diverse risk assessments, like all living beings? Will the tyrant concepts of health and normality finally lose their edge, making it clear that we are all at risk, and imperfect to a certain extent? The question of in what way to integrate (or not to integrate) predictive probabilistic knowledge into future generations' lifestyles is, of course, an additional problem.

Vogt and Green write convincingly that behavioural change is very hard to achieve in preventive, risk reducing medicine among adults. Do we wish to imagine a world in which babies are already disciplined and their behaviour medicalised in the hope of future risk prevention? Going back to the parents, it is argued repeatedly that knowledge is essential to allow pregnant women to make good, informed decisions about which tests to have performed during pregnancy, and what decisions to take in case of an abnormal diagnosis. Yet, is informed consent—that is, voluntary consent by a competent patient to whom full disclosure has been made, and who fully understands all that has been disclosed—still relevant to our times? Informed consent is one of the most praiseworthy ethical principles in contemporary medical practice. However, it is extremely hard to achieve in genomics or systems medicine in light of the huge amounts of difficult data for interpretation, and the fact that knowledge becomes ever more complex and probabilistic (Shkedi-Rafid and Hashiloni-Dolev 2018). Can even the most educated parents make sense of medical information that is probabilistic, futuristic and in fact, as described by Beck (1992) in his seminal work on risk society, often producing just more uncertainty, and as described by Vogt and Green, leading to over-diagnosis and often false positives? For the parents examining their foetus, this means a very complex cognitive, emotional, and moral experience. Thus I would suggest terming them not only moral pioneers, but educated moral gamblers, as parents understand that whatever their decision, it is not based on definite medical facts. That is because present day choice on continuation or termination of a specific pregnancy is sometimes based on an educated gamble (Hashiloni et al. 2019), which is itself grounded on probabilities, uncertainties, future risk assessments and predictions, intuitions, hopes, anxieties and faith.

As before, such complex and difficult decisions have to be made during the delicate phase of pregnancy, a stressful physical and emotional situation, and with time pressure, since pregnancies advance day by day. From the mothers' and fathers' points of view, the old challenges do not lose their relevance at all. What has enlarged their ability to choose, while at the same time narrowed their ability to accept their children just the way they are, has not changed. Nevertheless, choices have in certain cases become more difficult, futuristic, probabilistic and uncertain. The majority of future parents, living in developed societies that advance prenatal diagnosis, who do not resist testing completely, face even harsher dilemmas. That is because more knowledge does not necessarily bring more certainty. Often, genetic information is accompanied by VUS, uncertainty, false positives, over-diagnosis, gambling, and as a result—in the field of prenatal diagnosis—over-treatment, which is a larger potential for the termination of 'normal' pregnancies.

References

Beck, U. (1992). *Risk Society*. London: Sage.

Hashiloni-Dolev, Y., Nov Klaiman, T., Raz, A. (2019). Pandora's pregnancy: NIPT, CMA and WGS—A new era for prenatal genetic testing. *Prenatal Diagnosis, 39*(10), 859–864.

Katz-Rothman, B. (1986). *The Tentative Pregnancy: How Amniocentesis Changes the Experience of Motherhood*. New York: W. W. Norton.

Rapp, R. (1999). *Testing Women, Testing the Fetus: The Social Impact of Amniocentesis in America*. New York: Routledge.

Shkedi-Rafid, S., & Hashiloni-Dolev, Y. (2018). Pregnant genetic counselors in an era of advanced genomic tests: What do the experts test prenatally? *The Journal of Genetic Counselling, 27*, 1167–1174.

Part II

Philosophy of Biology

4

Developmental Narratives: How We Think that Organisms Use Genetic and Epigenetic Information

Christoph Rehmann-Sutter

The stories we tell about our biological development and about the action of the genes in our bodies contribute to the narrative construction of our identities. They contain views about who we think we are, albeit in a special way.

Of course, they do not belong to biographical narratives, since they are not recounting things we have experienced. And they are not recounting things that we did. They are not first-person narratives at all, because we did not make our biological development, nor did we experience it as we experienced those things we tell about in life-stories. The development of our bodies happened and continues to happen to us. Only exceptionally was it, or is it, a result of our own actions. Development began to take place before we even knew we existed, beginning with the zygote and going through embryonic stages, birth, and early childhood. And it goes on.

C. Rehmann-Sutter (✉)
Institute for History of Medicine and Science Studies, University of Lübeck, Lübeck, Germany
e-mail: christoph.rehmannsutter@uni-luebeck.de

D. Mahr and M. von Arx (eds.), *De-Sequencing*, Health, Technology and Society,
https://doi.org/10.1007/978-981-15-7728-4_4

Stories about development do not answer what qualitative psychologists call *classical identity questions*: 'What kind of a person am I, as what kind of a person do I wish my interaction partner to see and treat me?' (Lucius-Hoene and Deppermann 2004: 168). But they are none the less, as I shall argue, quite intimately connected to questions of identity, but on a deeper level: at the level of embodiment. Meaningful identity questions might be: 'What kind of living beings are we as humans? How do we believe our bodies are positioned in relation to the world, to others, to the events that happened to us, and to the genes we have inherited from our parents?' Storytelling is communicative and deeply practical. More precisely, it is a corporeal social practice—as Theodore Schatzki has put it—a connection between what we do and say ('nexus of doings and sayings', Schatzki 2001). Narrating bodies therefore belongs to what Annemarie Mol and John Law (2004) have aptly called 'doing the body'. It is doing the body with words and imagination.

4.1 Climbing the DNA

That gene-stories must have implications for our self-understanding became strikingly obvious to me nearly 20 years ago, when my family visited the Lawrence Hall of Science in Berkeley, California. It is a science museum, particularly well equipped for families with children. Our two daughters were 4 and 6 at the time. On the museum's roof terrace, overlooking San Francisco Bay, is a playground. By the side of a big grey plastic whale was a large, multicoloured climbing structure in the shape of the DNA double helix. Parents with a moment to spare could read a plaque on the side, with an explanation of what their children were climbing on. It read: 'DNA is the molecule that told your body how to become exactly who you are and how to stay alive. It is a long list of instructions written in a special code.' Wow! This was in 1997.

When I went back to there a few years ago, I saw that the plaque was still in place. I am not sure how many biologists today would still agree with this explanation. The plaque was designed before the completion of large-scale genome projects, and before epigenetics became widely discussed. It very much echoes how genetics was explained in the early

1960s: as a genetic program that instructs the cells of the body how to behave and how to shape the multicellular organism. That was before developmental genetics were able to investigate in detail how genes work during development. An accurate explanation of DNA today would instead place it within an interactive system that uses DNA sequences as one important information resource—but not the only one—in development. But in the public sphere and the media, the narrative of DNA as the blueprint of every being continues to resonate.

The explanation given to parents on the plaque is an interpretation of what DNA is. There is a complex mixture of biological facts combined with a high-octane cultural interpretation in these two sentences. It is a story that tells much more than the facts discovered by molecular biology could ever tell. It is about the meaning of DNA molecules for our lives.[1] And it has a temporal structure: DNA *told* your body is in the past, and to *stay alive* points to the future. It is a story about our development from the zygote until the end of our days. DNA tells the body during development exactly how to be shaped and organised, how to continue living, and what kind of persons are embodied by it. This is quite a lot. And that little story also describes how DNA does this job: as a long list of instructions, written in the genetic code. The body deciphers the instructions and then executes their message. It is about a molecule that controls our development by giving instructions to the body. We are what we are, because the DNA *told* it to our bodies. This is the message we get; and while parents are being told this DNA story, right before their eyes their children are climbing on that DNA, even as their bodies are repeatedly told by the DNA bodies exactly who to become. This is an example of what I shall call developmental narratives.

Developmental narratives are not *biographical narratives* (Robinson 2012), because development and embodiment underlie our lives, which are the subject of biographical stories. Developmental narratives are not about *narrative development*, nor are they about the development of a child's narrative capabilities, or about the key stories that play a role in the course of a child's life. I also use the term *developmental narrative*

[1] See Nelkin and Lindee (1995) and Rothman (2001).

in a different sense from child psychologists, who investigate the narratives of the experiencing child as the child tells them (Nelson 2010), or forensic psychologists who use narrative interview techniques to elicit reliable information from children; for example, to evaluate suspicions of child abuse, or in criminal court cases (Saywitz and Camparo 2014). My understanding of the term *developmental narrative* covers the ways in which we understand, evaluate and tell the story of embodiment from the zygote until the end of life. It covers scientific research fields such as embryology, developmental genetics, ageing research and gerontology, but it includes more. Developmental narratives are not purely scientific explanations, because they contain life-world meanings. These meanings are about identities, and they are social. They link us to how we became what we are. Today, how we became what we are is, in one way or another, explained in terms of genetics. Developmental narratives are therefore to be considered as scientific-cultural co-constructions of understanding embodiment.

As we can see in the example of the DNA climbing structure, the story contains three different characters: (i) the DNA, (ii) the body, which is the organism, and (iii) *you*, i.e. we ourselves who are the attentive audience of the storytelling. In other variants, however, (iv) the cells appear as further actors, as do (v) biomolecules other than DNA, (vi) the environment, and sometimes also (vii) the life-history of the individual. In some narratives (as is the case in this one), DNA is the only really active entity, and the others are passive executors of instructions, sometimes responding to selected cues from elsewhere in order to establish how the instructions are to be implemented. This sounds rather undemocratic. Other versions (for example, developmental systems theory; Oyama 2000), contain less centralistic and more democratic images of dispersed and joint cellular decision-making.

What is striking in developmental narratives is that most of them contain a story *within* the story, whose author appears as one of the figures within the play: the genetic instructions, the genetic information, the genetic program or the like, whose author is (in the quoted version) DNA, or in other versions a plurality of authors contributing to the continuous emergence of a developmentally relevant genetic information set. Hence, an analysis of developmental narratives needs to consider several authorships who appear in different roles: (i) we are the ones telling a story about (ii) DNA, the genes, and the genome which instructs the cell, or, in the more democratic variants, about the genetic information being read by the developmental system, in the context of (iii) a culture that infuses storytelling with a plurality of narrative frameworks, expectations, and the whole tradition of genetic tales, some of them more dominant than others.

Among the many questions that can be asked about developmental narratives, there is one on which I would like to focus in this chapter: Is there a bridge that relates a developmental narrative to our understanding of embodied life? If developmental narratives have implications for identity, and if concepts of embodiment have implications for our self-understanding, then the potential existence of such a bridge is not unlikely. Developmental lives and biographical lives are not completely separated, because they are not told independently of each other.

One potential bridging concept appearing in the recent literature is the notion of responsivity. In current reformulations of developmental narratives, which are inspired by epigenetics and systemic thinking, processes of development, i.e. the body subject, are described as 'responsive'. And responsivity is also part of how we understand the moral subject.

4.2 Responsive Minds in Responsive Bodies

This needs explanation. I start with the role of embodiment in the philosophy of practice. In a recent text about the role of embodiment in moral theory, Diana Tietjens Meyers states: 'The moral subject is not a disembodied reasoner conjoined to a disembodied will that issues commands to the body they're contingently housed in. Rather, the moral subject is an embodied perceptual, reflective, and volitional system that is alive to and responsive to the flux of events' (Meyers 2016). Meyers paraphrases (and agrees with) Aristotle's view of the moral subject, which she finds in his Nichomachean Ethics. Note the mention of *responsivity*. Further into her text she explains the proximity of this view to the phenomenological philosophy of Maurice Merleau-Ponty. She finds ample reasons to defend her view, which she calls 'psychocorporeal practical intelligence', against attempts to explain the mind–body relationship in more materialist, instrumentalist, or mechanistic terms. The key idea is that 'well-ordered choice and action integrate mind and body'.[2] We do not have an intelligent and responsive mind on the one hand and a body on the other, which is used by the mind to articulate responsive behaviour in reality. Responsivity, Meyers explains, belongs to our embodied presence both in the social word and within ourselves.

Presence has a past. Living things develop and reproduce. We became what we are now in the course of a complicated development history that once started with an embryo. Embryonic development, embodied growth during childhood, and ageing are processes that are not attributable to our agency as subjects. They happen within our bodies. Or better: it is our bodies that exist through these events. Modern biology, in particular embryology and genetics, have tackled these processes and captured them under the general term *development*. Our present scientific view of development is thoroughly geneticised. It is the genes that somehow account for the precise way our bodies develop. And *genes* now mean *DNA* and *genetic information.*

One might ask: Is there a connection between the mind–body relationship on the one hand and the relationship between living beings and their developmental information on the other? I think this may be one of the most important questions in the philosophy of biology. In living our lives, we are bodily present. We feel, see, communicate, and behave in correspondence with what is around us. We are responsively present. Is this capability of embodiment to be responsively present brought about by a sophisticated body machinery which is itself a product of a built-in genetic program? Or, as recent developmental genetic and epigenetic research seems to suggest, is development itself to be thought of as an essentially responsive process? And if so, what is organic responsiveness in relation to subjective responsiveness?

This is the question I want to address in the remainder of this chapter. I will not add any new empirical evidence. As a philosopher,

---

[2]Introduction to her chapter 'Psychocorporeity, the Autonomous Will, and Living Organ Donation', published in German as Meyers (2016). I quote from her original manuscript.

I depend on information gathered by molecular and developmental biologists in their experimental research. My task is to interpret, to conjecture, and to look for possible meanings, without drawing too hasty conclusions. Philosophy also needs to be critically attentive to the presuppositions that underlie certain scientific research practices and their popular explanations.

4.3 The Concept of *Epigenetics*

Let us now take a closer look at the term *epigenetics*. Following recent trends in the scientific literature, we understand epigenetics as the study of non-DNA sequence-based structural adaptations in certain chromosomal regions that change and co-determine the activity pattern of DNA. An epigenetic process is a cellular procedure that brings about such adaptations, changes them and stabilises them. They appear in cell lines during differentiation into a multicellular organism, and some of them are also transmitted from one generation to the next.

Epigenetics is relevant to the philosophy of biology because it seems to undermine some views of the way development is steered by genetic information. The central dogma of molecular biology, as formulated by Francis Crick in 1958, is well known. It assumes a linear correspondence between the sequence of three types of macromolecules in living cells: the nucleotide sequence of DNA and RNA, and the amino acid sequence of proteins. The dogma provides that in cells, there is a transformation of information from nucleic acid (DNA, RNA) to protein, but 'once information has passed into protein it cannot get out again'.[3] The flow of information is from DNA to RNA to protein and not the reverse. The embodied reality of the cell is largely made up of proteins; there are structural proteins (building blocks) and enzymes (biocatalysts), which can steer all kinds of biochemical reactions crucial to metabolism. In other words: genotype determines phenotype, and there is no information flow from the phenotype into the inheritable genotype.

[3]Crick (1958, pp. 152ff.), quoted after Griffiths and Stotz (2013, p. 40).

Put in extreme terms, as Richard Dawkins did in his famous evolutionary parable of *selfish genes* (Dawkins 1976), it seemed that DNA builds and uses the organism in order to survive in the world. The organisms are the *survival machines* of their genes. In using its organism, it is looking for an ecological niche it can exploit in order to multiply itself. Epigenetics contradicts this picture. It is not genocentric but applies contextual hermeneutics: instead of asking how organisms are used by their genes, epigenetics explores how organisms use their DNA in a proper and adapted way. Epigenetics seems to tell the story of a dependent DNA, used selectively by cells as an information resource in responsive ways that are appropriate within and adapted to an organismic and ecological context.

The term *epigenetics* was first introduced in 1942 by developmental biologist Conrad Waddington to refer to the study of causal interactions between genes and their products, i.e. the processes by which the genotype gives rise to the phenotype (Jablonka and Lamb 2002). The technical term was a fusion between *epigenesis* (the opposite of *preformation*) and *genetics*. Today, however, we call this line of research developmental genetics, or just developmental biology: it is the study of the processes of constructing a new organism from all inherited material and informational resources, in interaction with the environment. What contemporary *epigenetics* does is something else, with the *epi-* understood in a different sense: referring to causal factors in addition to the genome. Such causal factors can be *found* (medico-scientifically), and some can be *inherited*.

Most of them are not present at the zygote or the earliest embryo stages, but are acquired during the life of the organism. In humans, epigenetic influences begin during foetal development in the womb (Gunnarsson Payne 2016). And if some acquired information becomes inheritable, we see examples of so-called Lamarckian inheritance. Darwinism, as it appears in the shape of the *Modern Synthesis* in twentieth-century molecular biology, however, assumes that no acquired information is inherited. This is one of the main reasons why epigenetics has long been controversial. In the second decade of the twenty-first century, however, it has become scientifically respected, with papers now appearing regularly in leading journals such as *Nature*, *Cell*, or *Science*.

Research is ongoing while the available evidence is still limited. Miska and Ferguson-Smith (2016) concede that 'we do not currently have the complete picture for any natural example. Nevertheless, diverse organisms provide clues and candidate mechanisms', including human beings and other mammals.

Epigenetics can have either a predominantly developmental perspective (explaining how differentiation of cells in complex organisms takes place, and how development works), or a predominantly intergenerational one (explaining how inheritance works).[4] Within the genomic paradigm that focuses on chromosomes and genes, predominant in contemporary molecular biological research, it seems useful to define *epigenetics* narrowly as the study of structural adaptations in certain chromosomal regions that are not based in changes in DNA sequence: adaptations by which alterations in the activity states of the genome are registered, signalled, or perpetuated (Bird 2007). The best-known epigenetic systems currently under study are: (i) The Polycomb and Trithorax proteins, named after mutants of the fruit fly Drosophila, which bind to DNA and work to maintain certain genes down-regulated and to activate others; (ii) The methylation of selected cytosine bases of the DNA, cytosine being one of the four nucleotides that make up the DNA molecule; (iii) DNA is wrapped around histone molecules to build the nucleosomes. There are several post-transcriptional modifications of these histone proteins, including hyperacetylation and hypoacetylation, which affect gene activity within the affected genome regions; sometimes this is called the *histone code*; and (iv) There are also diverse RNA-associated regulatory systems.

In contrast to earlier proposals, for example by the group led by Arthur Riggs (Russo et al. 1996, relating back to Nanney 1958), this proposed taxonomy by Adrian Bird does not restrict *epigenetic* events to modifications that are inheritable. In theory, it could be that a regulatory intervention remains stable only over the lifetime of one cell within a tissue, which then dies without producing further daughter cells. From a developmental biology perspective, therefore, inheritance

[4]For a slightly different terminological classification, see Griffiths and Stotz (2013, p. 112) and Jablonka and Lamb (2002).

cannot be an absolute prerequisite. However, Bird's definition excludes other phenomena, such as the three-dimensional architectural patterns of cellular membrane systems, prions, or alternative self-sustaining, regulatory, metabolic feedback loops, which also might account for development. On the other hand, Eva Jablonka and Marion Lamb (Gissis and Jablonka 2011; Jablonka 2013) include them in their broader concept of epigenetics. They are interested in expanding the range of factors contained in predecessor cells beyond only the genotype. They are collecting examples of factors that all, in several different ways, influence the structure and properties of daughter cells, or daughter generations. In this perspective, the genome and all that is connected to DNA becomes one of several informational resources for development, but not the only one.

A picture is emerging of the DNA of cells being regulated in each single cell and cell line on-site, depending on the cell's changing local circumstances. The regulation of genome activity can respond appropriately to environmental stimuli, depending on the history of the organism's previous growth, its development and ageing, and remembering previous regulatory states. In a functionally ordered way, cells seem to react mutually to each other's presence by altering the regulatory states of their genomes. They are mutually responsive and reactive, and memorise the previous course of events. This is how embryonic differentiation is thought to take place, and how specialised tissues and organs emerge during development. By similar mechanisms, the organism also registers events that take place within its milieu. This emerging picture, drawn from the available cumulated evidence, is independent of the definition of the term *epigenetic* that we use in the context of certain experimental practices, i.e. whether we include self-maintaining regulatory loops and the architectural structure of the cell's membrane system, as do Jablonka and Lamb, or whether epigenetic research should focus primarily on chromosomal marking, as Bird has suggested. The picture is remarkably different from the earlier picture of organisms, which in the 1960s were imagined to be genetically programmed by their genomes.

There are many interesting implications of this—apart from the obvious responsiveness of the epigenetically organised developmental system (which I shall discuss in the next section). One is the phenotypic nature of the regulation of genotypic information. 'Epigenetics, in a broad sense, is a bridge between genotype and phenotype – a phenomenon that changes the final outcome of a locus or chromosome without changing the underlying DNA sequence' (Goldberg et al. 2007). Wilhelm Johannsen introduced the term 'genotype' around 1909 to counter those who believed in the inheritance of characteristics. He stressed that what is inherited is not a set of characteristics but a set of factors that determine—influenced by the environment—the way the plant grows from the seed. It is not the phenotype, which is the actual plant, that is inherited, but only the genotype. The genotype is what is contained in the zygote that can explain, in part, the features of the developing offspring (Griffiths and Stotz 2013: 13). Therefore, all the cells of a multicellular organism share the same genotype. And cellular differentiation (producing the cellular phenotype) may be considered an epigenetic phenomenon, largely governed by stable and partly inheritable changes in gene expression. The black box between genotype and phenotype, which stood between geneticists and embryologists for much of the twentieth century (Peirson 2012), is wide open for detailed research.

Second, the study of inheritable epigenetic modifications would also meet Johannsen's original definition of the term *gene*: *something* in the gametes that conditions or co-determines a property of the progeny (Johannsen 1909: 124). For good reason, Johannsen did not stipulate any material definition of what this 'something' might be. The gene in Johannsen's sense is therefore not restricted to DNA. Inheritable epigenetic factors (*epigenes*) could also be considered genes. Genetics is not preoccupied with studying DNA and its effects alone but also with all the factors in the gametes that account for the emerging characteristics in the next generation. Why not include the regulatory marking of DNA in the concept of the genome? Epigenetics then merges with genetics. The idea of genes and genomes would expand and include additional things alongside DNA; epigenetics would become a redundant term. Johannsen's definition of a gene does not distinguish genetics from epigenetics.

If we want to keep this distinction, we need an additional feature. According to current use, the *genome* is an organism's complete set of

DNA, including all of its genes.[5] The genome thus includes the genes but is more than the complete set of genes; it also includes the DNA regions that have regulatory functions beyond *coding regions* for polypeptides. But it does not include the epigenetic patterns. We can refer to coding and de-coding that occurs in the DNA-related processes of transcription and translation, but not in chromatin marking and all the other processes that are counted under epigenetics. The characteristic of the genetic system of inheritance, in contrast to the epigenetic system, can thus be coding. Epigenetic factors are therefore not included in the genome because they are not coded.

4.4 Organic Responsivity

Within this emerging *epigenetic* picture, the genome is not a centralised source of instructions with which it steers developmental processes. Instead, it is an information resource that can be activated and used by the cells in a way that corresponds to the situation of the cell within its surroundings, and depends on the previous developmental history of the organism (Shapiro 2009). Niewöhner (2011) calls it the 'molecularisation of biography and milieu'. The developmental system, which is represented by the organism and its genetic and epigenetic processes, is in several crucial ways 'responsive' to the environment. It seems to have a built-in openness both to neighbouring cells within a tissue and to the environment around the organism.

This is not entirely new. In hindsight, we can recognise that a responsive capacity was already part of older models, long before epigenetic research in its contemporary sense took off. Griffiths and Stotz (2013: 50ff.), for example, pointed out that the *lac operon* model of gene regulation, developed by François Jacob and Jacques Monod in the early 1960s for *E. coli* bacteria, was open to environmental influences: regulatory molecules (such as repressor proteins), together with the corresponding regulatory sites of structural genes, made up feedback loops through

[5] See, for example, the NIH Genetics Home Reference: What is a genome? https://ghr.nlm.nih.gov/primer/hgp/genome, accessed 3 July 2019.

which environmental cues could influence gene regulation. Systemic responsiveness is therefore not restricted to epigenetic processes, but it seems to be the main task of the cell's chromatin marking systems to make the activity patterns of its genome responsive. Its regulatory state is specific to a cell with a distinct interactive, developmental history.

Adrian Bird (2007) uses the word *responsive* to characterise the role of epigenetic marks. They are not proactive, but respond to things that have already taken place: 'An implicit feature of this proposed definition is that it portrays epigenetic marks as responsive, not proactive. In other words, epigenetic systems of this kind would not, under normal circumstances, initiate a change of state at a particular locus but would register a change already imposed by other events.' Responsivity means being capable of registering and remembering. But then he continues: 'It could be argued that the responsive nature of epigenetic processes is a unifying feature, because classic epigenetic systems such as the DNA methylation system and the Polycomb/Trithorax systems seem to respond to previous switches in gene activity in this way. Therefore, their sophisticated feature is the ability, in the *darkness* of the nucleus, to sense and mark changes in the chromosomal status.'

This *sensing* should, of course, not be conflated with subjective perception. Bird does not depict epigenetic processes as having experience of any kind. It is just a way of registering patterns. In the same paragraph he explains: 'Without such epigenetic mechanisms, hard-won changes in genetic programming could be dissipated and lost.' He continues to describe the cell within the frame of macromolecular biochemical machinery that is genetically programmed. The program cannot, however, be there from the beginning of development like a film roll that is spooled off. The program is modified during development. *Hard-won* changes in programming need to be memorised, which can be accomplished through epigenetic processes. The program in Bird's sense is therefore a developmental potential for the cell which, while connected indirectly to the stable DNA, itself has a history of being influenced and changed by previous developmental events.

Eva Jablonka (2013) uses the word *responsive* in an evolutionary context. She quotes empirical findings where phenotypic adjustments to environmental challenges that persist for generations are followed by genetic assimilation and stabilised, as 'conditional ontogenetic responsiveness'. In addition, 'changes that increase responsiveness' can be part of an evolutionary process, which can then 'start with phenotypic rather than genotypic change'. Even the germline may be part of it, as she suggests: 'The germline has evolved to be a particularly active cell type with mechanisms that ensure genome functionality and variability.' Transposition (i.e. the change of positions of genes in chromosomes) can play a role in the germline, in order to 'increase the range of inducible, selectable, epigenetic variations in descendants'. The germline can contain epigenetic variations, which are acquired, and are not all deleted in gametogenesis: 'Traces of epigenetic variations in the form of variant small RNAs, variant templating proteins, or variant methylation or histone traces, which are determined by the parental germline, enable the reconstitution (probabilistically) of the parental variant state (e.g. of DNA methylation).'

Gissis and Jablonka (2011: 406 ff.) suggest that the *Modern Synthesis* of Darwinist evolutionary theory and molecular and population genetics must be fundamentally rewritten. But epigenetics does not claim that the sequence of DNA is inherited with a Lamarckian logic; rather, epigenetic inheritance should be viewed as 'soft inheritance' (Haig 2007), it works at the level of how the DNA is used by the cells. Epigenetic adaptation capacities considerably enhance organisms' fitness. Organisms are no longer be understood as the 'survival machines of their genes', as Richard Dawkins (1976) had described them within a strictly genocentric framework.

Chromatin marking by epigenetic processes joins a series of other well-known phenomena where developmentally relevant 'genetic' information is not read directly from the DNA sequence but is constituted in actual developmental contexts. Alternative splicing is another of these phenomena. Cells can construct several different mRNA templates for protein synthesis from the same DNA gene, by cutting out different parts as *introns* and combining the remaining *exons*, depending on the actual needs of the cell. Another phenomenon is mRNA editing, where the mRNA sequence is changed in the cytoplasm before it is used for protein synthesis, in order to produce the right protein.[6]

One theoretical framework that can accommodate the multifunctionality of DNA and epigenetics is the developmental systems view, as introduced by Susan Oyama (1985). This assumes that developmentally relevant genetic information is not inherited by the next generation as a package of instructions for making the structure of the adult. In sharp criticism of any genetic representationalism contained in the notion of a genetic program, or the 'list of instructions', she writes: 'What is transmitted is macromolecular form, which, though it is necessary for the development of phenotypic form, neither contains nor constitutes plans for it, and developmentally relevant aspects of the world' (Oyama 1985: 22). Information is not conveyed to the body cells in the shape of instructions on how to develop. The information content of a *developmental system* itself has a history within development. The developmental system can expand within a certain niche. Hence, also certain aspects of the world belong to inheritance.

The key tenets of a developmental systems approach have been listed (Oyama et al. 2001): (1) joint determination by multiple interactants; (2) context sensitivity and contingency, in the sense that the significance of any one causal element depends on the state of the rest of the system; (3) extended inheritance; (4) development seen as a process of reconstruction of traits, which are not inherited as such; (5) no single type of interactant controls development; (6) evolution of organism-environment systems that change over time. The developmental systems approach is therefore a wholeheartedly epigenetic approach to development (Griffiths and Hochman 2015).

In this view, the developmentally relevant information is not a script, which is there from the beginning, but information is itself constructed step by step by the system-in-an-environment, in a sequence of events that build on each other, always through interaction with selected

[6]See Griffiths and Stotz (2013), Schmidt (2013), Shapiro (2009), Neumann-Held and Rehmann-Sutter (2006), and Rehmann-Sutter (2002).

bits of DNA. Information, as Oyama puts it, has an 'ontogeny' in a developmental process.[7]

Theoretical interpretations of epigenetics, such as the developmental systems approach, and other versions that have emerged around the idea of responsiveness, give rise to different developmental narratives. I believe that they have some important philosophical implications, because they change significantly the image that we have of our own embodiment. With all necessary caution, I would phrase it as follows: embodiment, on all its ontological levels—from the molecules to cells to the developing organism to relationships and sociality—is a responsive process of some kind. Responsiveness does not appear only after the machinery of the body is completed, i.e. on the upper level of psychocorporeality. It is no less than an organising principle of all development.

However, caution is necessary to avoid drawing conclusions too rapidly. In my interpretation of organic responsivity I do not go as far as Le Chapellier and Matta (2011), who claim that 'cells have probably a psychic part which is an elementary material unconscious' (p. 173). The idea of a cellular psyche or of a material unconscious is troubling, since the idea of the unconscious presupposes consciousness. And it conflates conscious perception with the sensitivity of cells to certain aspects of their environment, and their capacity to adapt responsively and to stabilise certain adaptations. The latter is on the level of macromolecular form, of cellular structure, of biochemical reactions, which need certain genes to be active and to activate others. If *receptors* on the surface of a cell are involved, these are also macromolecule complexes that can bind to a certain type of molecule and then trigger reactions to it. Therefore I cannot believe that there is an unconscious in a cell that would justify the talk of *cellular perception* in a psychologically relevant way. The philosophical implications of developmental narratives, in my view, are on a different level. I would rather follow James Shapiro's suggestion of an 'informatic viewpoint' (Shapiro 2009).

[7] For a comment on this view, see Rehmann-Sutter (2008: 42–45).

4.5 Corporeal Presence

Is there a continuity of organic responsivity and psychocorporeal responsivity? Or are there breaks, different ontological levels of responsivity to be considered? How does the term 'response', which is taken from communication, change its meaning when we apply it at the level of cellular epigenetic processes? Can the concept of 'response' be stretched so far? And how can the semiotic transitions involved be explained? The few remarks I can make here are intended to invite further discussion rather than close these questions.

One important discontinuity is subjectivity. Epigenetic processes are not subjective, because there is no experiencing subject present in them. This is important to keep in mind, in order not to give away a strong sense of agency that is necessary for ethical responsivity. However, the caution goes even further: the development of our bodies from the embryonic stages throughout life is neither a subconscious nor an unconscious process. Consciousness and its related terms (sub- and un-) are not the right category to capture it. The processes of development are beneath, or more fundamental, than any possibility of an organism being conscious or unconscious. Development, I would prefer to say, is *a-conscious*, or *a-subjective*.

There are many different ontological levels involved in human embodiment, which could perhaps be distinguished as follows: (i) the a-subjective but responsive development; (ii) the crypto-subjectivity and subconsciousness of some embodied processes—such as the need for bonding or the development of sexual desire; (iii) the unconscious in a Freudian sense as a result of repression and defence; (iv) the consciously experiencing of subjectivity; and finally (v) reflecting and moral subjectivity. On all these levels, there is a kind of responsivity, but the stimuli and the responses are quite different on the different levels.

The discussion of these levels should not be conflated with the problem of animal subjectivity. Thomas Nagel's famous question, 'What is it like to be a bat?' (Nagel 1974) is a meaningful question, not despite but precisely because we cannot know as humans, who lack ultrasound radar and cannot fly, how it is to be a bat. We can say: it must be somehow for the bat to be a bat. Or it must be somehow for an octopus to be an octopus. When we see a bat flying, we appreciate that there is a possible perceptive world there which we cannot know: the world from the bat's point of view, the 'bat's world'. Bats are sensing and behaving animals. Bats, or octopuses, have some form of agency within their own behavioural worlds, but I cannot know much of it. Their perception relies on their sensory organs and is connected to innate patterns of behaviour. Hence the appreciation of difference is the key to a description of animal subjectivity that does not do injustice to the animals. Animal subjectivity is not a deficient version of human subjectivity, but it is different from human subjectivity, and certainly also different from species to species.

We cannot insert other life forms—plants and animals—into an ontological scheme that stems from a description of human embodiment (such as the five ontological levels I have just loosely sketched out). It reminds of the old theory of different levels of the soul: the vegetative, the sensitive, and the rational soul, all three of which happen to be present in humans. Animals and plants are not described fairly using a deficiency model, which starts from a human self-description: animals are those lacking rational souls, plants are those lacking sensitivity. Such schemes justify the superiority of our own species, and do not do justice to other species.

But perhaps there is an important continuity in the word *respond*. Maurice Merleau-Ponty, in his *Phenomenology of Perception* (*Books I/III*) (1962), has described the corporeal movements of lived bodies as continuous presence in both space and time. In order to say this, he relies on his experience as a lived body. Is development also a process that can aptly be described as continuous presence in both space and time? Responsiveness implies some form of being present: the cellular developmental system is on an informational level present in the context of other cells, and so on.

The phenomenological perspective Merleau-Ponty adopted is tied to the subjective point of view of a human experiencing subject. As such, we can ask: what is our lived experience of embodiment? How is the body that I am given to me as the perceiver of my body? My body is my way of being present in the world: sensing and making movements, being positioned relative to others, using body knowledge and so forth. The body, in this *embodiment phenomenological perspective* (Van Manen 2016: 127–131), is not an object that appears before my eyes. I do not have my body. I am embodied. I am in relation to the world. The body is a kind of in-between. I am not a living being, or an organism, Merleau-Ponty emphatically declares, in order to make us adopt the radically subjective and first-person experiential perspective of my (or your) lived body, which is alien to the 'organism'. The words *living being*, or *organism* do not fit into the experience of embodiment. Of course, I know that I am a living being and an organism, but this is a different perspective from lived experience. I live my body by being bodily present in space and time, in relation to things around me, living and non-living.

Therefore, I believe that the theories that describe and explain the development of an organic system, including developmental systems theory, epigenetics, and genetics, do not describe or explain my lived experience of being embodied in the world. Developmental theories are presented as knowledge. They give rise to an imagination of how my body developed before I became conscious, to an imagination of how my body, the body as which I am present in the world, develops further. On this level of imagination, a developmental narrative can make a difference, because this imagination contains a developing body (which I know that I am), that is embodied present to the world even at the deepest level of cellular differentiation. Each cell is oriented to *other*, in order to be able to develop.

Corporeal presence has two levels, which are therefore more closely related to each other than the 'DNA-as-a-list-of-instructions' narrative would suggest: development and behaviour or moral subjectivity are all ontologically different, but they are all, in one way or another, responsive. None of them is a self-contained way of functioning, otherwise the other levels would not be possible.

One implication of this continuity and discontinuity of presence, if carefully considered, is that in order to see clearly the bridge between body phenomenology and the (epi-)genetics of development through the lens of *responsivity*, we need to translate genetic knowledge from its objective scientific context (epigenetic responsivity) to the dramatically different epistemic situation of lifeworld and corporeality (psychocorporeal responsivity). If genetic and epigenetic information are to become meaningful in the practical contexts of our lives, we need to develop skills to dissect developmental narratives into layers of meaning. We need a capability to deconstruct critically and understand what happens in the course of such translations. Developmental narratives, as they come, might not be explicit enough about them.

References

Bird, A. (2007). Perceptions of epigenetics. *Nature, 447,* 396–398.

Crick, F. H. (1958). On protein synthesis. *Symposia of the Society for Experimental Biology, 12,* 138–163.

Dawkins, R. (1976). *The Selfish Gene*. Oxford: Oxford University Press.

Gissis, S., & Jablonka, E. (2011). Final discussion. In S. Gissis & E. Jablonka (Eds.), *Transformations of Lamarckism: From Subtle Fluids to Molecular Biology* (pp. 395–409). Cambridge MA: MIT Press.

Goldberg, A. D., Allis, C. D., & Bernstein, E. (2007). Epigenetics: A landscape takes shape. *Cell, 128,* 635–638.

Griffiths, P., & Stotz, K. (2013). *Genetics and Philosophy: An Introduction.* Cambridge: Cambridge University Press.

Griffiths, P. E., & Hochman, A. (2015). *Developmental Systems Theory*. eLS. Chichester: Wiley. https://doi.org/10.1002/9780470015902.a0003452.pub2.

Gunnarsson Payne, J. (2016). Grammars of kinship: Biological motherhood and assisted reproduction in the age of Epigenetics. *Signs: Journal of Women in Culture and Society, 41,* 483–506.

Haig, D. (2007). Weismann Rules! OK? Epigenetics and the Lamarckian temptation. *Biology and Philosophy, 22,* 415–428.

Jablonka, E., & Lamb, M. J. (2002). The changing concept of epigenetics. *Annals of the New York Academy of Sciences, 981,* 82–96.

Jablonka, E. (2013). Epigenetic inheritance and plasticity: The responsive germline. *Progress in Biophysics and Molecular Biology, 111,* 99–107.

Johannsen, W. (1909). *Elemente der exakten Erblichkeitslehre.* Jena: G. Fischer.

Le Chapellier, P., & Matta, B. (2011). Cellular perception: When the cell model includes a sense order which ensues from a philosophy of nature, the signaling and epigenetics effects which can result from exposure to magnetic fields are described better. *Neuroscience & Medicine, 2,* 161–177.

Lucius-Hoene, G., & Deppermann, A. (2004). Narrative Identität und Positionierung. Gesprächsforschung. Online-*Zeitschrift zur verbalen Interaktion, 5*, 166–183. ISSN 1617-183. www.gespraechsforschung-ozs.de. Accessed 12 August 2019.

Merleau-Ponty, M. (1962). *Phenomenology of Perception* (Trans. C. Smith). London: Routledge.

Meyers, D. T. (2016). Psychokorporeität, autonomer Wille und Lebendorganspende. In T. Moos, C. Rehmann-Sutter, & C. Schües (Eds.), *Randzonen des Willens. Anthropologische und ethische Probleme von Entscheidungen in Grenzsituationen* (pp. 113–131). Frankfurt a. M: Lang

Miska, E. A., & Ferguson-Smith, A. C. (2016). Transgenerational inheritance: Models and mechanisms of non-DNA sequence-based inheritance. *Science, 354*, 59–63.

Mol, A., & Law, J. (2004). Embodied action, enacted bodies: The example of hypoglycaemia. *Body & Society, 10*, 43–62.

Nagel, T. (1974). What is it like to be a bat? *Philosophical Review, 83*, 435–450.

Nanney, D. L. (1958). Epigenetic control systems. *PNAS, 44*, 712.

Nelkin, D., & Lindee, S. (1995). *The DNA Mystique: The Gene as a Cultural Icon*. Freeman: New York.

Nelson, K. (2010). Developmental narratives of the experiencing child. *Child Developmental Perspectives, 41*, 42–47.

Neumann-Held, E. M., & Rehmann-Sutter, C. (Eds.). (2006). *Genes in Development: Re-Reading the Molecular Paradigm*. Durham, NC: Duke University Press.

Niewöhner, J. (2011). Epigenetics: Embedded bodies and the molecularization of biography and milieu. *BioSocieties, 6*(3), 279–298.

Oyama, S. (1985). *The Ontogeny of Information: Developmental Systems and Evolution*. Cambridge: Cambridge University Press.

Oyama, S. (2000). Causal democracy and causal contributions in developmental systems theory. *Philosophy of Science, 67*(Proceedings): S332–S347.

Oyama, S., Griffiths, P. E., & Gray, R. (2001). Introduction: What is Developmental Systems Theory? In S. Oyama, P. E. Griffiths, & R. D. Gray (Eds.), *Cycles of Contingency: Developmental Systems and Evolution* (pp. 1–11). Cambridge, MA: MIT Press.

Peirson, B. R. E. (2012). *Wilhelm Johannsen's Genotype–Phenotype Distinction*. Embryo Project Encyclopedia. ISSN: 1940-5030. http://embryo.asu.edu/handle/10776/4206.

Rehmann-Sutter, C. (2002). Genetics, embodiment and identity. In Grunwald et al. (Eds.), *On Human Nature: Anthropological, Biological, and Philosophical Foundations* (pp. 25–50). Berlin: Springer.

Rehmann-Sutter, C. (2008). Genetics, a practical anthropology. In M. Düwell, C. Rehmann-Sutter, & D. Mieth (Eds.), *The Contingent Nature of Life: Bioethics and the Limits of Human Existence* (pp. 37–52). Berlin: Springer.

Robinson, M. (2012). The biographical narrative in popular culture, media and communication: An introduction. *Networking Knowledge: Journal of the MeCCSA Postgraduate Network, 5*. https://ojs.meccsa.org.uk/index.php/netknow/article/download/289/123/. Accessed 11 August 2019.

Rothman, B. K. (2001). *The Book of Life: A Personal and Ethical Guide to Race, Normality, and the Implications of the Human Genome Project*. Boston, MA: Beacon.

Russo, V. E. A., Martienssen, R. A., & Riggs, A. D. (Eds.). (1996). *Epigenetic Mechanisms of Gene Regulation*. Woodbury, NY: Cold Spring Harbor Laboratory Press.

Saywitz, K. J., & Camparo, L. B. (2014). *Evidence-Based Child Forensic Interviewing: The Developmental Narrative Elaboration Interview*. Oxford University Press: Oxford.

Schatzki, T. (2001). Practice mind-ed orders. In T. R. Schatzki, K. Knorr-Cetina, & E. von Savigny (Eds.), *The Practice Turn in Contemporary Theory* (pp. 50–63). London and New York: Routledge.

Schmidt, K. (2013). *Was sind Gene nicht? Über die Grenzen des biologischen Essentialismus*. Bielefeld, Germany: Transcript.

Shapiro, J. A. (2009). Revisiting the Central Dogma in the 21st Century. *Annals of the New York Academy of Sciences, 1178*, 6–28.

Van Manen, M. (2016). *Phenomenology of Practice: Meaning-Giving Methods in Phenomenological Research and Writing*. London: Routledge.

Photo 2 © Grischa Lichtenberger, Grischa Lichtenberger, 0815_11_sc_dia uni b_30

5

Epigenetics, Responsiveness and Embodiment

Maria Kronfeldner

I would like to comment on the connections Rehmann-Sutter (Chapter 4) establishes between epigenetics, responsiveness, and embodiment.

Epigenetics is hype and despite the limited evidence available, it has solidified a new conception of DNA as 'responsive', and rightfully so. This new conception of DNA has gradually built up over the 'century of the gene,' to use Keller's (2010) phrase. Even though epigenetics is now a buzzword that is used in at least five senses (see Nicoglou and Merlin 2017), I do not want to discuss the pros and cons of the different usages, since the differences do not matter much for the context at issue.

Comment on Ch. Rehmann-Sutter 'Developmental narratives: How we think that organisms use genetic and epigenetic information'.
To appear in *De-sequencing. Identity Work with Genes*, ed. by D. Mahr and M. v. Arx.

M. Kronfeldner (✉)
Central European
University, Vienna, Austria
e-mail: Kronfelderm@ceu.edu

D. Mahr and M. von Arx (eds.), *De-Sequencing*, Health, Technology and Society,
https://doi.org/10.1007/978-981-15-7728-4_5

After introducing one such usage as a foundation for the discussion, I will discuss three issues that concern subsequent claims about epigenetics. I will come back to a different usage of the term 'epigenetics' at the very end of discussing the third issue.

5.1 Epigenetics?

For the following, I will assume the meaning of epigenetics that Rehmann-Sutter suggests, which more or less is already present in the work of Jablonka and Lamb (2005). It includes, in contrast to other interpretations, 'structural adaptations in certain chromosomal regions that change and co-determine the activity pattern of DNA' and 'three-dimensional architectural patterns of cellular membrane systems, prions, or alternative self-sustaining, regulatory, metabolic feedback loops'. The core of epigenetic processes is that they regulate gene action without changing the DNA. DNA is in that sense 'responsive' to its epigenetic 'regulators.' Basically, the language of responsiveness and regulatory processes means that DNA can react to affordances that result from the interaction with the environment.

I will not discuss this characterization itself; rather, I want to comment on three issues related to how it is used in Rehmann-Sutter's chapter—issues related to the ways in which epigenetics, responsiveness, and embodiment are connected.

5.2 Responsiveness as Agency or Causal Process?

First, we should be careful (hence the quotation marks in the above characterization of epigenetics) with *any* intentional, agency-suggesting language for molecular processes. It is either metaphorical language or it needs much more detailed justification of why it does not reduce to ordinary causal processes that—in the age of epigenetics—include processes of gene regulation.

Rehmann-Sutter is aware of the danger in such language when he distances himself from any claims about a cellular psyche or material consciousness, and when he states that one needs to understand the change in meaning of the term 'response' if it is applied across the five 'ontological levels involved in human embodiment'. With respect to this layering of what 'response' means at these different levels, he claims, for example, that non-human animals and humans have different forms of subjectivity. Yet, given what he writes, I still wonder what the responsiveness of the first, the molecular 'a-subjective' level amounts to, if not sheer causal interaction, as in any other causal process in the world? Certainly, these causal processes are 'present in the context of other cells' but that *presence* is a far cry from the presence that enters a phenomenological 'experience *as* lived body'.

Thus, the first issue I want to raise is the following: is there any difference between saying that there is an ongoing causal interaction of all kinds of developmental resources in parity (i.e. no causal primacy of one over the other) and saying that there is a 'form of agency' at that first ontological level of human embodiment? I doubt it. Given that the history of talking about these things is, ever since Aristotle, laden with terms such as 'agency', 'power' and the like, as stressed in particular by developmental systems theory, parsimony should be the guide in the way we talk about genetics, epigenetics, and development.

5.3 Epigenetics as Reason for Responsiveness at All Levels?

Rehmann-Sutter claims that epigenetics and its various theoretical interpretations 'have important philosophical implications, because they change significantly the image that we have of our own embodiment. Embodiment, on all its ontological levels—from the molecules to cells to the developing organism—is a responsive process. Responsiveness does not appear only after the machinery of the body is completed, i.e. on the upper level of psychocorporeality. It is the organising principle of all development.' The conclusion—responsiveness is everywhere—is certainly correct (if we take, for the sake of argument, the language to be neutral with respect to the first issue mentioned above). It even, one feels tempted to add, applies to *evolution all the way down*, if philosophers such as Campbell (1974) or Dennett (1995) are correct, defending that there are basic processes of cognition very early in life (in Campbell's case) and 'kinds of minds' (in Dennett's sense) equally early. Rehmann-Sutter also points to that evolutionary level when he mentions Jablonka's (2013) use of responsiveness. The problem is, however, that this then means that responsiveness in epigenetics is no big news. Responsiveness at the most basic ontological level is neither stemming from epigenetics, nor is it original or specific to epigenetics; it goes all the way back to the beginning of life. Rehmann-Sutter is aware of this when he writes that the lac operon model of gene regulation was already pointing to 'systemic responsiveness' which is 'therefore not restricted to epigenetic processes'. Yet, if the latter is the case, then that is in tension with the above claim about the philosophical importance of epigenetics.

Thus, the claim that should be defended with respect to responsiveness, taking all of the above into account, is that responsiveness (in the basic sense that can be defended even at the molecular level) is a property of *life*—whichever, i.e. a property of any life, be it life in our actual world with epigenetics or a life in a hypothetical (and maybe early world of evolution) without epigenetics. Epigenetics might further confirm that picture of overall responsiveness, but epigenetics is not the reason why there is responsiveness all the way down.

The *specific* philosophical importance of epigenetics is thus more limited. As Rehmann-Sutter mentions, it revises the so-called central dogma by allowing some reverse 'information flow' and it turns Dawkins (1976) 'selfish gene' picture upside-down: for Dawkins, DNA uses organisms; epigenetics means that 'organisms use their DNA'. An exaggeration, since sometimes the epigenetics *happens to* the organism (e.g. via environmental stress) as the DNA happens to the organism, but an exaggeration that rightfully corrects another exaggeration. (One might go on to add further specific issues for which epigenetics has been taken to be relevant—a frequently mentioned one is that it allegedly reintroduces Lamarckism—but the point I want to make here is independent of any such list of specific implications.) Important here is that the claim that epigenetics turns the selfish gene picture upside down is a much more specific claim than the claim that epigenetics shows that there is responsiveness all the way down to basic molecules.

As a matter of parsimony again, I think one should stick with specific claims or refer to the actual reason why there is responsiveness,—namely, because we are talking about life.

5.4 Embodiment Already in Nature–Nurture Interaction

Finally, because both Chapter 4 and the volume are about how identity and genes connect, be it via narratives or other ways of relating to one's lived body, I also want to raise some doubts about the tight connection Rehmann-Sutter makes between embodiment and epigenetics. Ever since nature versus nurture became an issue, embodiment has been an issue too. And nature versus nurture is a very old problem, going back to at least Greek antiquity and thus did not, as is often claimed, begin only in the nineteenth century with Francis Galton (Kronfeldner 2018a). Certainly, any mind–body dualism is deciding the issue from the very start, entailing the irrelevance of nature (equated with the body). Yet not all nature–nurture discussions were framed in a dualist mind–body ontology hypothesizing a mere body as distinct from a disembodied mind, and definitely not the ones in philosophy of biology, spearheaded by biologists such as Richard Lewontin. In addition, there are those scientists, e.g. Caspi et al. (2002), who show empirically that nature (in the context of their research: genetic and other innate factors) and nurture (which includes lived experience) interact in a strong sense—in a sense where it depends crucially on the environment in which the individuals live (abusive or not), as to *which effect* the respective genes have on the individual, e.g. increasing (one effect) or decreasing (another effect) the probability of aggressive behaviour. If the data accumulated in this field are tracing causal processes—from an environment via individuals (with their bodies, including the brain) to the behaviour of these individuals—, then these individuals literally *embody* the abusive or non-abusive environment (see Kronfeldner 2018b for more on nature–nurture interaction). And these individuals will have quite some stories (i.e. narratives) to tell about that embodiment.

To the best of my knowledge, it is not yet clear whether the causal process that would account for the data accumulated by Caspi et al. (2002) and others involves epigenetic processes, i.e. changes in gene regulation (for a review, see Tabery 2014). But it is clear, and independent of whether the process in fact involves epigenetics, that the gene–environment interaction at issue is a case of embodiment. It *will still be a case of embodiment even if it turns out that it does not involve any changes in gene regulation.* One could certainly—by definitional fiat—say that epigenetics is even more inclusive than as introduced above, including a*ll* processes where an interaction of the body with an environment changes the 'response' of the body (i.e. the behaviour of the individual). Such a *very* broad concept of epigenetics (mentioned briefly when Rehmann-Sutter refers to Goldberg et al. 2007) is close to useless since, on the basis of such a very broad concept, even the causal processes that account for why a plant grows taller when it receives more water would *by definition* be an epigenetic process, since 'the final outcome of a locus or chromosome' is changed 'without changing the underlying DNA sequence'.

My third point is thus: the embodiment Rehmann-Sutter discusses is not dependent on epigenetics as long as epigenetics does not include everything, i.e. any gene–environment interaction, which will include almost all processes of development (i.e. all those that do not follow a flat norm of reaction, since only in a flat norm of reaction, can the gene have the same effect irrespective of the environment in which the body lives).

The risk in using a very broad concept of epigenetics (including almost everything as epigenetics) is that the discussion about development continues in a polarised manner: with a straw-man of genetic determination (i.e. a flat norm of reaction) put in opposition to a too-broadly-defined concept of epigenetics, with nothing in between.

5.5 Summary

Rehmann-Sutter points to important aspects of how genetics, epigenetics, responsiveness, and embodiment are connected, as part of his account of developmental narratives. Yet, what he writes could be (mis-)read as still too laden with agency-talk and as mislocating the source for responsiveness and embodiment.

References

Campbell, D. T. (1974). Evolutionary epistemology. In G. Radnitzky & W. W. Bartley (Eds.), *Evolutionary Epistemology: Rationality and the Sociology of Knowledge* (pp. pp. 47–89). LaSalle, IL: Open Court Publishing.

Caspi, A., McClay, J., Moffitt, T. E., Mill, J., Martin, J., Craig, I. W., et al. (2002). Role of genotype in the cycle of violence in maltreated children. *Science, 297,* 851–854.

Dawkins, R. (1976). *The Selfish Gene*. Oxford: Oxford University Press.

Dennett, D. C. (1995). *Darwin's Dangerous Idea: Evolution and the Meanings of Life*. New York: Simon & Schuster.

Goldberg, A. D., Allis, C. D., & Bernstein, E. (2007). Epigenetics: A landscape takes shape. *Cell, 128,* 635–638.

Jablonka, E. (2013). Epigenetic inheritance and plasticity: The responsive germline. *Progress in Biophysics and Molecular Biology, 111,* 99–107.

Jablonka, E., & Lamb, M. J. (2005). *Evolution in Four Dimensions: Genetic, Epigenetic, Behavioral and Symbolic Variation in the History of Life*. Cambridge, MA: MIT Press.

Keller, E. F. (2010). *The Mirage of a Space Between Nature and Nurture*. Durham, NC: Duke University Press.

Kronfeldner, M. (2018a). Divide and conquer: The authority of nature and why we disagree about human nature. In E. Hannon & T. Lewens (Eds.), *Why We Disagree About Human Nature* (pp. 186–206). Oxford: Oxford University Press.

Kronfeldner, M. (2018b). *What's Left of Human Nature? A Post-essentialist, Pluralist, and Interactive Account of a Contested Concept*. Cambridge: MIT Press.

Nicoglou, A., & Merlin, F. (2017). Epigenetics: A way to bridge the gap between biological fields. *Studies in History and Philosophy of Science Part C: Studies in History and Philosophy of Biological and Biomedical Sciences, 66*(December), 73–82.

Tabery, J. (2014). *Beyond Versus: The Struggle to Understand the Interaction of Nature and Nurture*. Cambridge: MIT Press.

6

Space and Time of Developmental Narratives

Claudine Burton-Jeangros

Christoph Rehmann-Sutter's chapter focused on interactions between developmental narratives and identity providing an opportunity to discuss connections between biological lives and social ones. While genetic determinism long prevailed, assuming that genes are given and inalterable, recent developments in epigenetics are now suggesting that they are influenced by exposures to environmental circumstances, opening up new research avenues across disciplines (Landecker and Panofsky 2013). The articulation of responsive minds and responsive bodies, expanded into *moral subjects*, direct attention to individual trajectories as they develop in situated social contexts. In this short sociological commentary, I would like to further the argument, based on Maurice Merleau-Ponty, that lived bodies are situated in both time and space. In fact, I consider that taking into account time and space, as elements relating the experience and narratives of individuals to their socio-historical context, is crucial in the understanding of the many

C. Burton-Jeangros (✉)
Institute of Sociological Research, University of Geneva, Geneva, Switzerland
e-mail: Claudine.Jeangros@unige.ch

D. Mahr and M. von Arx (eds.), *De-Sequencing*, Health, Technology and Society,
https://doi.org/10.1007/978-981-15-7728-4_6

layers of meaning in which Rehmann-Sutter is interested. Space and time define social, cultural, economic, political, and scientific conditions that unquestionably affect developmental narratives, through available resources, including forms of knowledge on what development means. While the genome has gained a lot of credit, it is indeed only one among 'several informational resources for development'.

The importance of the context will be illustrated with an *extreme* example, namely the impact of the Chernobyl accident on the lives of those who were exposed to nuclear radiation after the catastrophe. Adriana Petryna (2002) coined the concept of *biological citizenship* to refer to the community created by a shared damaged biology, a condition on which affected citizens have claimed compensation benefits. Another relevant aspect of this case relates to the constant debate about the actual consequences of the exposure to radiation, with competing interpretations. While some, particularly those directly concerned, claimed evidence of physiological damage based on symptoms and illnesses they experienced after the accident, others, notably those in official positions at that time, only acknowledged a possible psychosocial impact in terms of stress and fear, without accepting any physiological influence. Beyond the biological impact of this technological catastrophe, as a source of exposure to damaging effects, the political transition in the aftermath of the accident from the Soviet regime to the independent state of Ukraine, provided to the latter an opportunity to claim national autonomy. To me, this extreme situation shows the importance of unpacking the many layers that any developmental narrative entails, including, beyond individuals perceptions, the influence of a range of social and scientific institutions. These institutions also shape developmental narratives in mundane trajectories, even if they certainly do it in a less visible way.

Next to spatial embedding in a social context, the lifecourse perspective that developed in the 1960s (Elder 1999) provides relevant insights to take into account the temporal aspect of developmental narratives. While aiming at understanding individual lives as a lifelong process from birth to death, this approach also invites the situating of trajectories and individual ageing into their historical time, thus examining variations across cohorts. Again, material and symbolic resources change over time, affecting the type of developmental narratives that may be told and possibly producing dissension across generations. More specifically, lifecourse epidemiology added a temporal dimension in the analysis of health inequalities, including a focus on embodiment describing 'how extrinsic factors experienced at different life stages are inscribed into an individual's body functions or structures' (Kuh et al. 2003: 780). Systematic variations in health conditions reveal how a range of social determinants impact individuals' physiology differently, with adverse conditions resulting in less optimal development and poorer health over the lifecourse for some, a process also observed in high income societies where the overall level of health is elevated. Nowadays, epigenetics adds a layer in the understanding of the complex articulation between molecular and societal levels, with the observation that exposure to specific socio-economic conditions, in sensitive developmental periods, is stored in cells, and possibly transmitted across generations, contributing to the reproduction of social and health inequalities (Vineis et al. 2015).

Taking into consideration space and time or, put differently, context and period, allows us to further the idea of *responsivity* defined in Rehmann-Sutter's chapter as the capacity to register and remember, by considering other influences affecting the interactions between responsive minds and responsive bodies, thus adding to the complexity of the analysed mechanisms. Next to understanding, recent knowledge in epigenetics also encompasses policy implications along with a new economy of hope based on potential action on genes (Rose and Novas 2003). Ideas about shifting biologies can sustain individual and collective strategies to anticipate and mitigate some risk exposures for the sake of good health (Landecker and Panofsky 2013). Consequently, it is also important to examine how institutions and social arrangements can favour strategies that either reduce or exacerbate social inequalities. This is an ambitious agenda, made possible by interdisciplinary research.

References

Elder, G. H. (1999). *Children of the Great Depression: Social Change in Life Experience* (25th anniversary ed.). Boulder, CO: Westview Press.

Kuh, D., Ben-Schlomo, Y., Lynch, J., Hallqvist, J., & Power, C. (2003). Life course epidemiology. *Journal of Epidemiology and Community Health, 57,* 778–783. https://doi.org/10.1136/jech.57.10.778.

Landecker, H., & Panofsky, A. (2013). From social structure to gene regulation, and back: A critical introduction to environmental epigenetics for sociology. *Annual Review of Sociology, 39,* 333–357. https://doi.org/10.1146/annurev-soc-071312-145707.

Petryna, A. (2002). *Life Exposed: Biological Citizens After Chernobyl*. Princeton, NJ: Princeton University Press.

Rose, N., & Novas, C. (2003). Biological citizenship. In A. Ong & S. Collier (Eds.), *Global Anthropology*. London: Blackwell.

Vineis, P., Kelly-Irving, M., Rappaport, S., & Stringhini, S. (2015). The biological embedding of social differences in ageing trajectories. *Journal of Epidemiology and Community Health*. https://doi.org/10.1136/jech-2015-206089.

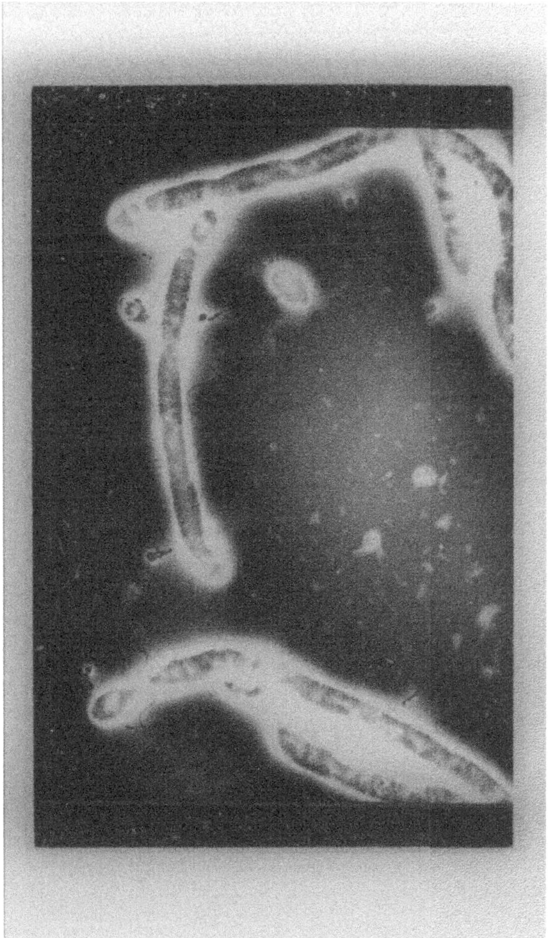

Photo 3 © Grischa Lichtenberger, Grischa Lichtenberger, 0815_11_sc_dia uni b_27

Part III

Societal Contexts

7

Data Mining in Systems Medicine and the Project of Solidarity: The Interface of Genomics and Society Revisited

Barbara Prainsack

7.1 Introduction: From Genomic Medicine to Precision Medicine

In 2013, the American author Dave Eggers wrote a book titled *The Circle*, in which he presented a dystopian fantasy of a future in which people's lives take place primarily through social media. *The Circle* is the name of the social media company that bears a striking resemblance to Facebook, except that its penetration into all aspects of people's lives is even more pronounced than in the case of its actual-world counterpart. In the book, people have forgotten what it is like to live offline, and

This chapter draws in part upon the following book: Barbara Prainsack. 2017. *Personalized Medicine: Empowered Patients in the 21st Century?* New York: New York University Press.

B. Prainsack (✉)
Department of Political Science, University of Vienna, Vienna, Austria
e-mail: barbara.prainsack@univie.ac.at

Department of Global Health & Social Medicine, King's College London, London, UK

D. Mahr and M. von Arx (eds.), *De-Sequencing*, Health, Technology and Society,
https://doi.org/10.1007/978-981-15-7728-4_7

anybody who tries to reconnect to their old analogous selves is cast out of society, or, in the worst case, even killed in an attempt to *help* them. They are little more than their digital data doubles, consisting of social media activities and other digital traces. Being *social*—not in the old-fashioned sense of engaging with other people, but rather living in and on social media—has also taken over politics: in Dave Eggers' story, the social media market is dominated by one big company that ultimately also takes over the role of running elections.

Medicine, of course, features very prominently in the book. One would expect that healthcare has turned into a 'Gattaca'-like[1] enterprise where people have their genome sequenced at birth and are stratified into different groups according to their genetic predispositions; perhaps those with very poor genetic predispositions will be sorted out right away so that society saves itself the cost of investing in them. What we find in the book, however, is very different. In none of the scenes that revolve around healthcare, and there are quite a few of them, is DNA even mentioned. The scenarios are futuristic: in the first scene, Mae, the protagonist of the story, who has just started a job at *The Circle*, is called in for an appointment with an occupational physician. When Mae arrives, the physician already knows more about Mae's lifestyle and health status than does Mae herself, mainly through Mae's health information that is stored online, and through other things that can be learned from analysing Mae's social media activities. During her appointment with the doctor, a number of tests are run on her, and she is asked to drink a smoothie containing nanosensors; these will measure a number of biomarkers from inside her body. She is also given a cardio-sensor to wear around her wrist. Every two weeks, Mae is told, she will need to come back to see her doctor, who will run a standard battery of tests including blood work, cognitive tests, reflexes, eye exams, and occasionally also MRIs. There is no mention of genes or genomes.

What is happening here? In the book, at least, the answer is simple: using other, non-genetic information is cheaper, and, even more importantly, it is more actionable. The type of medicine that is featured in the

[1] 'Gattaca' is the title of a 1997 movie by Andrew Niccol that portrays a dystopian future of genetic determinism. It has since become a shorthand term for societal visions of genetic essentialism and genetic determinism.

book is all about prevention. When Mac enquires about the cost of the comprehensive monitoring that her company provides for her, free of charge, her doctor explains:

> Mae couldn't figure it out. 'But how is this affordable for you guys? I mean, the cost of an MRI alone –'
>
> 'Well, prevention is cheap. Especially compared to finding some Stage-4 lump when we could have found it at Stage 1. And the cost differential is profound. Because Circlers are generally young and healthy, our health care costs are a fraction of those at a similar-sized company – one without the same kind of foresight.' (Eggers 2013: 153)

How does this futuristic fictional scenario live up to the actual, non-fictional world? In real life, hardly anybody has their DNA sequence on file. It is also unlikely, I believe, that newborns will soon be sequenced at birth and their information stored somewhere central—if only because the storage costs are too high. The cheapest way to store DNA information is still in the human beings themselves; whenever genetic or genomic information is needed, this can be accessed on an ad hoc basis, both cheaply and rapidly, allowing healthcare professionals to zoom into the exact markers and information that a specific diagnostic or therapeutic question requires. It can be done in the doctor's office, or even in the patient's own home, with portable DNA sequencers; the data can be shared with the patient's healthcare team via a secure online tool, or if the patient so desires, even with researchers more broadly. The tools for doing the latter already exist today, and I shall return to them shortly. With regard to other types of monitoring, the scenario in *The Circle* is not so far removed from what is done today. Admittedly, even in the richest countries, people are not given nanosensors to swallow. But some are asked to wear a rubber bangle that measures their heart rate (e.g. Baig et al. 2013). Others are expected to monitor their own health information via an online platform—and apps such as Apple Health Kit on mobile phones and smart watches are making it ever easier to do so.

In 2011, the National Academy of Sciences in the United States published a report titled *Toward Precision Medicine* (NAS 2011). This was a time when researchers and policymakers in many countries across the globe thought of ways to utilise the growing availability of digital data to *personalise* medicine, in the sense of tailoring medicine more closely to the specific characteristics of individual patients. What this report suggested was quite revolutionary. It proposed to radically change the way we think about health and disease. Our current disease taxonomy is symptom-based, and it has outlived its usefulness. Now that we know more and more about individual differences, we also know that no disease manifests itself in exactly the same way in any two people. My genes, my diet, my age, my life circumstances, the drugs that I am taking, and the conditions that I suffer from, are all factors that influence how my migraines present themselves. In any case, my migraines are not the same as anybody else's migraines, and lumping us together into one disease group leads to problematic healthcare decisions: another person may need a totally different treatment than I do.

The awareness that people suffering from a similar disease may need a different treatment is, of course, not new: this is exactly the type of expertise that health practitioners who know their patients and their families have been bringing to the table for decades... and centuries. What has changed, however, is what type of information is collected and used for this purpose. The *evidence* used needs to be *hard*, digital (or at least digitisable), and computable (McMahon et al. 2020; Prainsack 2020). The US-based organisation Autism Speaks, for example, uses a range of biomarkers including DNA testing and brain imaging to stratify people with autism spectrum disorders into smaller groups that may benefit from the same kind of therapeutic intervention. Also, the National Academy of Sciences (NAS) wanted to capture individual differences between people in a very systematic and particular way: they used the basic idea of the Google Maps feature as a template to think about how different types of information could be combined to create a map of a person's health or disease at any moment in time. Some of the data in this map would be very, or relatively, stable, such as the genomic environmental or environmental exposures, whereas others would be very dynamic, such as disease symptoms and activity scores (activity trackers are not depicted in this map—but remember that this Report was published before activity trackers became popular). In short, the

National Academy of Sciences advocated a scenario where old symptom-based disease taxonomies were replaced by data-rich representations of individuals in different stages of health and disease. Such a new disease taxonomy, they argued, would bridge the needs of the clinic and the needs of research (which is converging more and more in any case), and it would support healthcare decision-making that is tailored much more *precisely* to the individual characteristics of patients.

In this vision of precision medicine, genomics is no longer a soloist; it is part of a much larger orchestra. And the National Academy of Sciences' Report is not alone in taking this approach; there was a large Report on Personalised Medicine by the European Science Foundation (ESF) that defined Personalised Medicine broadly as the consideration of individual characteristics in every stage of medicine, from prevention, diagnosis, treatment, to monitoring (ESF 2012). Many authors envisage the inclusion of wide ranges of data, including non-molecular data such as data from people's wearable or remote sensors. Some authors envisage the inclusion not only of your personal health records and fitness tracking data from your smart phones or sports bangles (if you use those items), but also your social media activity, your criminal records, your credit card purchases (e.g. Weber et al. 2014). There is so much data out there, so the argument usually goes, and we have a moral duty to use it.

What does this mean for ELSA (an acronym meaning studying ethical, legal, and social aspects of) genomics? I argue that there are a number of assumptions that are made in such visions of personalised medical futures that people who study genomics and medicine in society are well placed to scrutinise. I shall now discuss two of the assumptions that I believe we need to unpack.

7.2 Citizens as Health Altruists: Is There a Societal Duty to Collect More Data?

The first assumption is that there is a societal or moral, or even a legal duty to collect and use more data. Some people argue there is a moral duty to use data from people's personal domains to *personalise* their healthcare. The language in consent forms, information sheets, and online platforms often suggests that people who do not allow others to use their data are holding back the development of social goods. Accusing citizens who refuse to grant access to their personal or health data of being selfish or ignorant is part of the strategy of some companies who use the *sharing* rhetoric to recruit more customers, such as the direct-to-consumer genome testing company 23andMe famously did. What renders this assumption problematic is that it involves a slippage from the societal responsibility and accountability we should expect from corporate actors to what we should expect from individual citizens. Large corporations that collect and use data—whether a biobank, a public agency, or a private company—do arguably have a societal duty to consider if the datasets in their possession could be put to use to create public benefits. Very often these corporations use the data they obtain from citizens at no cost to serve their own purposes—to monitor people more effectively, to patent a new genetic marker, or to please their shareholders. Against this backdrop it is reasonable to expect that they give something back to society if this is reasonably and meaningfully possible. But the same does not apply to citizens: citizens who protect information and data about themselves often do so to protect themselves from harm or worry. They do so because they know that the organisations using their data are much more powerful than *they* are, and that citizens have little if any redress if they are harmed by what the organisations do.

In this context, conveying to people that handing over their personal health data is an act of altruism, or even their moral duty as a citizen, is part of a discourse where the idea that profits can be private and losses and risks should be borne collectively has become hegemonical. Such calls on citizens to be data altruists when many corporate actors are data hoarders are not articulated only by commercial actors who want to get their hands on more personal data. Also, patients undergoing clinical treatments in publicly funded hospitals are asked to make their tissue and information available for research in medical establishments, and many non-profit websites and platforms are now available that provide the infrastructures for sharing genomic and other data. What makes this landscape of citizen health altruism so difficult to navigate is that it is hard to separate platforms and organisations who call upon citizens to 'share' their data for purely self-serving reasons from those who foster genuinely solidaristic ways of sharing with the aim of creating public benefits without harming the interests of individuals or groups (examples are dna.land or openhumans.org; see also Prainsack 2015).

This trend of calling upon citizens to share information about ever more aspects of health and their lives bears an uncanny resemblance to two of the slogans that the company in *The Circle* uses: 'Secrets are Lies! Sharing is Caring'. Also in the book, the purposes of sharing are portrayed as positive—it will help to improve health, to provide better entertainment, to allow a more 'honest' way of dealing with one another—while being silent about the side-effects, namely the loss of control by citizens of more and more areas of their lives while large, for-profit corporate actors gain ever more.

Ironically, this trend is supported by another, parallel development, which was driven by the best of intentions: this development is a change in the relationship between human rights and scientific research. For a long time, the relationship between the two was understood primarily in terms of a right of people to be protected from the negative consequences of science. In the medical domain this meant that people needed to be protected from things being done to them without their informed and voluntary consent. In the context of genomics, important human rights issues were protection from discrimination on the basis of genetic information, and the protection of the human species from interventions in the human germline. If we look at publications on the topic of human rights and genomics from around 2010, then those are the issues in the foreground. Today, this is changing. An increasing number of scholars and activists refer to Article 15 of the International Covenant on Economic, Social and Cultural Rights (1966), which defines the human right to 'enjoy the benefits of scientific progress and its applications'. As so often, the devil lies in the detail, namely in how far we intend the duties and obligations such a right imposes on some actors to go. If people have a right to enjoy the benefits of scientific research, does this mean that there is an obligation for research to 'open up' and use all available datasets in medical research institutions? Or do we also need to 'open' clinical data? Do people have a moral, if not a legal, obligation to allow others to use their personal health information if it will be used for research with a likely public benefit? And what does 'opening datasets' mean in this context: does it mean that we need to make these data available only to bona fide researchers, or to everybody?

We have only just begun to consider these questions. I believe that people working at the interface of medicine, genomics, and society are particularly well placed to help avoid the assumption that underlies many of the current activities to use and share health data, namely that a human right to benefit from science means that we need to collect more and more data. How do we do that? The solution does not lie in balancing an individual human right to benefit from science with an individual right to privacy. Instead, privacy needs to be seen as a collective concern at least as much as it should be seen as an individual right. We need to think, as a society, where we can make crucial differences in the lives of people by using other people's data for research and for healthcare decision-making, and where the commitment to a society in which nobody is made to feel guilty for not sharing their data outweighs this need. Because sharing is not always caring, and big data alone, as media studies scholar Gina Neff reminds as, will not cure us (Neff 2013).

One step towards this goal is to stop using the term *sharing* when we speak of data and instead speak of making data or information accessible to somebody for a specific purpose. This forces us to be very specific about who should have access to data and information, and for what purpose, and avoids our well-meant sharing efforts ending up benefiting not those we seek to empower but to help those who already have enough power.

7.3 Can Genomic and Other Data Easily Be Integrated and Rendered Meaningful?

The flawed conclusion that collecting more and more data and sharing them will solve problems brings me to the next assumption that underpins visions of personalised and precision medicine. The assumption consists of two parts: first, that data will be interoperable; and second, that it is worth spending large sums of money to make them interoperable. Regarding the first part of the assumption, let us look at the 'Google Map' of health data again: the difference between the real Google Maps feature some of us use on our computers or our smart phones on the one hand, and the vision of the Google Map of health data on the other is that in the former case, the data that are integrated are already there in a digital and computable format, and we know what they mean. We have collected census data as well as topographical and traffic information with specific purposes for use in mind, and have several decades and in some cases centuries of experience in collecting, using, and integrating such information (city maps and road maps are examples). This is not the case for the types of health data that the authors of the Google Maps of health data scenario seem to bring together. How DNA-based information of a person who has a familial predisposition to cardiovascular disease can be integrated with data from her activity tracker or her diet to yield meaningful, medically actionable information is a complicated question. Moreover, even data that represent the same things, such as data in electronic health records, are typically not interoperable, because the companies providing the systems have a vested interest in locking users into their proprietary systems and software rather than enabling them to use their products in combination with products of other companies (Wachter 2015). Policy-makers in many countries have recognised this as a problem and begun to provide financial incentives, as well as some 'sticks', to encourage healthcare and IT software providers to make systems more interoperable (ONC 2015).

A question that tends to be sidelined amid efforts to make datasets interoperable, is whether it is worth making them interoperable. Is the benefit worth the cost? One of the reasons that questions about the benefit not only of integrating, but also of collecting and analysing data do not even get asked in some situations lies in the big data enthusiasm that also underpins approaches in personalised and precision medicine. A 2013 book by internet governance experts Viktor Mayer-Schönberger and Kenneth Cukier is perhaps the starkest illustration of what some authors have called the big data craze (Mayer-Schönberger and Cukier 2013). Titled *Big Data: A Revolution that Will Transform How We Live, Work and Think*, the book posits that big data entails literally unprecedented opportunities to obtain new insights – because it enables us to mine datasets for correlations without knowing what is the nature of their relationship. This data mining approach is said to end the 'age-old search for causality' (Mayer-Schönberger and Cukier 2013: 14). Instead of starting with the exploration of causes, these authors argue, big data allow us to discover patterns and correlations that can lead to 'novel and invaluable insights ...If millions of electronic medical records reveal that cancer sufferers who take a certain combination of aspirin and orange juice see their disease go into remission, then the exact cause for the improvement in health may be less important than the fact that they lived' (Mayer-Schönberger and Cukier 2013: 14).

A second characteristic of the big data paradigm, besides its supposed potential to abolish causality, is a changing attitude towards the accuracy of individual measurements. For people working with small datasets, according to Mayer-Schönberger and Cukier, it is important for every single data point to be correct. For those working with big datasets, in contrast, the larger scale makes up for the inaccuracies in single data points so that the signal can still be located in the noise. If, for example, a person measured her temperature once a day, it would be important for each measurement to be accurate if she wanted to obtain a reliable fever curve over the course of the week. If she measured her temperature every three seconds, then it would be practically irrelevant if a few of these measurements were incorrect.

I believe that for scholars at the interface of genomics and society, the critical examination of the assumptions and consequences of such an approach in medicine is an important task. The big data approach as described by Mayer-Schönberger and Cukier may be suitable for marketing, where it does not matter so much if a company sends an advert for hair product to a bald person. In medicine, in contrast, the health and lives of people are at stake. Here, one of the problems with a big data approach is the conflation of correlation and causation. To stick with the Schönberger and Cukier's example of aspirin and orange juice, knowing that 80% of the cancer patients in a study benefited from this combination would not enable us to predict whether or not a particular patient will benefit from it. When the causes for a particular effect are not known, the utility of the information for individual patients is limited, regardless of the size of the dataset—because even if something has an effect on most people, it does not mean that it will have the same effect on everyone. Neither is it true that a larger scale of data can always compensate for inaccuracies at the level of individual data entries. If there is a problem with the recording of the data, then a larger scale of data will also scale up the inaccuracies (Spiegelhalter 2013).

Despite these concerns, Schönberger and Cukier's big data approach has had a big impact on how personalised and 'precision' medicine is envisaged. The U.S. Precision Medicine Initiative launched by President Obama is just one example. In the same year in which Obama announced the Initiative in his State of the Union Address (The White House 2015), the British Health Secretary at that time, Jeremy Hunt, gave a speech to reporters emphasising the importance of preparing now to for some of the biggest challenges healthcare has ever faced. But Hunt was not talking about ageing societies and the increase in patients who lived longer with a greater number of chronic health conditions. Nor did he refer to the spread of obesity or diabetes. No, he was talking about replacing doctors. Once we have successfully implemented precision medicine and big data, Hunt argued on returning from a trip to California, we shall no longer need human decision-makers, because computers can do it better.

> 'If you talk to technology gurus in California and ask what's going to change in the next two decades,' Hunt said, 'they say "in 20 years' time, no doctor will ever give a diagnosis". They say "You can get three hundred thousand biomarkers from a single drop of blood, so why would you depend on a human brain to calculate what that means when a computer can do it for you?"' I think it's really important that we're ready in the NHS to harness the power of data to give us more accurate diagnoses. (Matthews-King 2015)

What may seem no more than a tempting yet naïve way by a Conservative politician to spot an opportunity to cut costs, is, however, deeply problematic. Problematic because of the moral duty bestowed on people to share their data, and because relying on computers alone to make healthcare decisions is a problematic idea. First, it is shortsighted because it introduces a decision-maker into the clinic that is not accountable to anyone. Second, it is problematic because if it was up to the Jeremy Hunts of the world, we would be channelling money into building infrastructures for the data-driven medicine that is missing from frontline healthcare services. Moreover, such visions give much more power to commercial companies. Commercial companies are called into build the technical and digital infrastructures for such big data approaches, and data is sold to them. (Remember that the government's plan to share people's medical records with private companies was one of the reasons that the English Care.data initiative failed; Van Staa and Goldacre 2016). Sometimes the data are given to commercial companies because they promise to do something good with it. But even then, companies often collect more data than is needed for the task, as the example of the Google-owned start-up company Deep Mind showed in spring 2016: the company was tasked with developing algorithms to help identify children at risk of kidney failure. For this purpose they received access to patient data within several Primary Care Trusts in England, independent of whether the data in question were necessary for the task (Hodson 2016). The rationale to justify this is, very often that, in times of budget cuts, we need this money—or the know-how 'donated' by consumer tech companies—to do good things, even if it comes from a bad place.

Last, but not least, the 'Jeremy Hunt vision of personalised medicine' is problematic because it leads, if not programmatically then practically, to the evidence used for medical decision-making being limited to data that are available in digital and computable format, ejecting personal meaning and the human touch even more from the field of medical decision-making. But how do we resist such visions of personalised medicine, visions where genomic and other data are enlisted to treat people's digital data doubles in silico rather than focusing on high-touch treatment of real people in ways they consider meaningful?

7.4 Does Personalised Medicine Necessarily Lead to Narcissistic, Individualistic Medicine?

Donna Dickenson, in her book *Me Medicine v. We Medicine*, argues that personalised medicine also corresponds with, and in fact supports, a form of 'patient narcisissm' that comes at the cost of collective wellbeing. There are many great things about this book, in particular also how Dickenson draws attention to the ways in which the need for personalised medicine is an economic and political one as much as it is clinical and scientific. But I disagree with her on the idea that we need to choose between 'We Medicine' and 'Me Medicine', and that the very idea of personalised medicine entails a shift towards a kind of ruthless and atomistic individualism. I disagree with it for two reasons. The first is that I believe that the choice between 'We Medicine' and 'Me Medicine' rests on a false dichotomy. People's engagement with their health, including decision-making about their healthcare, is not a solipsistic, narcissistic process. Even in the area of direct-to-consumer (DTC) genetic testing, which is seen by many as the epitome of individual narcissism, empirical research has found that most people take the tests because they are curious to learn more about to whom they are related, or about their own health, so that they can discuss it with their family or consider how it might affect their offspring (e.g. Cherkas et al. 2010; Harris et al. 2016). Personal genomes—and this is, I believe, one of the most important lessons from ELSA Genomics since 2010—are both personal and social. And when I say that they are social I do not mean this in the sense that Paul Rabinow's notion of biosociality has been interpreted, namely that group identities will be based increasingly on genetic markers, and people will mingle with others with whom they share genomic characteristics in common (Rabinow 1992). What I mean is that the ways in which we use genomic information are deeply social. Genomic information means something only because we relate to others. If somebody is interested in improving their own health, in learning about their DNA, or accessing their medical records, we should not simply assume that this is another instance of, as Carina Dennis called, it 'the rise of the narcisseome' (Dennis 2012). More often than not, the fact that they want to do this, and how the information obtained in this way will be used, has to do with other people. Also, data protection experts recognise this and are thinking of ways to accommodate the fact that health information is personal not only to the person that it comes from. Graeme Laurie in Edinburgh and Mark Taylor in Sheffield have pursued this argument about genetic data, but it also applies to health related information and the practices of people (Laurie 2002; Taylor 2013). If medicine is becoming more personal this does not mean it is becoming more individual.

The second reason why I disagree with having to choose between 'We Medicine' and 'Me Medicine' is that I believe that calls for more 'We Medicine' overlook the great extent to which commercial interests have become part of it already. The biggest DNA collection in the world is held not by a government or a public institution, but by a private company, 23andMe, in California. Private companies provide the tools and infrastructures to sequence our genomes, and share our data. Private companies fund and provide platforms for the networking of patient organisations. Private companies also own the information we use on search engines when we look for symptoms, and our online profiles. Private enterprises build the activity trackers that we use, and own our data. The fact that consumer scoring is becoming such a big problem is just one illustration of this. Consumer scoring is the practice of commercial services looking for information about people and assign scores to them according, for example, to how likely they are to default on a mortgage, to be suffering from a chronic disease, or to become *frequent flyers*—meaning that they face repeated admissions to hospital emergency rooms. They sell this information to companies, such as banks or healthcare providers, to whom the information is highly useful. They say it will help in identifying patients who are in need of extra attention—but it does not take a lot of imagination to figure out what the flip side of the special treatment will be. The state—and this is a very important point in this context—is a facilitator in all this. To date, we have called for more state regulation to keep these practices in check, but this has not proved successful. State actors do not crack down on large corporations, mainly they enable. As Linsey McGoey (2015) showed in her book, even global health philanthropy is dominated by private companies, to whom state actors give grants to help the poor. When the European Commission ruled in 2016 that Starbucks and Fiat had not paid enough tax, it was national governments that protested. Let us repeat this: the national governments who were not paid enough tax by those companies protested when the EU said that this tax should be paid. The *raison d'etat*—the reason of state—has, quite literally, been replaced by the primate of business interests. Coming back to medicine, big corporations are increasingly replacing public actors as funders; I believe that governmentality has been replaced by economentality. In large parts of the world, this is the reality of 'We Medicine', and I am not sure I want that.

The way out of this is through us, the citizens. What I mean here is not a further individualisation of responsibility, but that working towards a kind of medicine that is accessible for all and contributes to decreasing, rather than increasing, health and social inequalities is a fundamental democratic issue. It needs to be a collective concern. It is an issue that should be at the core of the relationship between people and collective actors. In fact, much of the activism around health at present can be understood in these terms. The Open Access movement, for example, believes that findings from research that was supported by public money should be available to the public with as few barriers as possible, and certainly without cost. The Open Science movement goes one step further by arguing that because science is always part of society, it should be accessible to all who want to participate in it. This is one of the reasons why proponents of Open Science argue that, in some situations, research data and the processes of scientific knowledge creation should also be made open. The term *open*, here, I believe, is actually not a very good word, as it can mean everything and nothing. The opposite of *open* is not *closed*. If data or information sources are not open this can mean that they are proprietary, difficult to find, impossible to read, or too expensive. Moreover, openness is not always good; just as it is said that it's possible to kill someone with kindness, people can be harmed by too much openness.

Having said this, fighting for more openness is an important goal when it helps to mitigate some of the most harmful power asymmetries in society, when it helps to create greater reciprocity between more powerful and less powerful actors, between citizens and corporations, between research participants and research institutions, and between patients and health professionals. It is for this reason I believe that if people want to have access to the raw data that are stored about them in databases, they should be allowed to access these. Societal change will not come from the corporations that run databases and provide healthcare, including our current governments. Change will come from people who are democratic innovators in the field of health. Such democratic health innovation can take the form of a person donating all of their health data to research, as do the participants of the Personal Genome Project (personalgenomes.org). But it does not need to go that far. Such democratic innovation also comes from people who refuse to participate in things that are expected of them and thus challenge hegemonic assumptions. Today, when we hear the term *disruption*, we think of a new technology—but these people are also disruptors! Moreover, very important democratic innovators in the field of health are also those who strengthen solidarity-based arrangements. Solidarity, I believe, is an underrated concept that actually has huge potential to help us to address challenges at the crossroads of genomics and society, and personalised medicine more broadly.

There are a number of ways in which solidarity can help us to help us to change how we frame certain questions, and suggest new solutions. One reason why solidarity is so useful is that it overcomes the dichotomy between self-interested and other-regarding action, acknowledging that most of the things human beings do include elements of both (albeit not to the same extent). Solidarity is thus uniquely placed to help us think about how we can accommodate the fact that health data, for example, are personal to more than one person. It can also change, as we argue in the book, how we organise the governance of databases, which is one of the key infrastructures underpinning personalised medicine. One of the things solidarity brings to database governance is that individual control over data is not the overarching goal, and that we can, as Barbara Koenig eloquently put it, give our consent to be governed

(Koenig 2014). This does not mean that we give up or overrule personal autonomy, but to move from a liberal understanding of autonomy to a relational one. A solidarity-based approach also shifts our attention to things that people have in common, rather than what sets them apart. As we show in our book (Prainsack and Buyx 2017) this has consequences for what we think about excluding people from specific services and treatments in the context of personalised medicine. Foregrounding what people have in common over what sets them apart does not mean that we gloss over differences or pretend that differences do not exist, but we say that commonalities rather than differences should be what guides policy-making and practice.

Another, equally important, consequence of solidarity is that we cannot just suffice to say that we accept people's solidaristic contributions to the greater good without ensuring that society does not abandon people if things go wrong. This is why we have proposed the establishment of harm mitigation funds in every context where people's data are being used (McMahon et al. 2020; see also Prainsack and Buyx 2013). But ensuring that people are not deserted when things go wrong has a much bigger implication at the systemic level of society as a whole. Personalised medicine must never be an excuse for the retreat of public responsibility. We need to think very carefully about new patterns of exclusion that may result from innovation in medical science and technology. I am not talking about people who are excluded from medical treatment on the basis of a somatic marker that suggests they may not benefit from it but still suffer side effects, as this would be a *good* way of excluding someone. I'm also talking about data-rich precision medicine that requires that people contribute data and information, or monitoring efforts, online, which brings the digital divide much more into the foreground: those who cannot engage with these online and digital tools will probably suffer consequences if we do not take action now to pre-empt that.

Furthermore, we need to protect not only people who cannot, but also those who chose not to, participate. This is not only a futuristic scenario that appears in books such as *The Circle*. In 2015, the Californian drug company Amgen offered a significant cost reduction to patients (US$14,000 per year reduced to slightly over US$60 a year) for a cholesterol reducing drug. All the patients needed to do was to give permission to the company to access and use their personal data for ten years (Hiltzik 2015). The company emphasised that this was entirely voluntary: people could choose whether they opted for the cheaper drug and sign over their data, or pay the higher cost. (This is a little like telling people at an airport that they can freely choose whether to be searched at security: if they choose not to, they cannot board their plane, but it is really up to them.) As scholars studying the role of medical genomics in society, we can all help to avoid a situation where people who do not choose to be entirely transparent, who have their entire lives, literally and metaphorically sequenced and shared, are suffering negative consequences. Perhaps the most important way to do this has nothing to do with cutting edge technology, but with a basic shared human need: the need to receive adequate healthcare. In an era of genomic and personalised medicine, making an argument for publicly funded healthcare accessible to everyone is more important than ever.

References

Baig, M. M., Gholamhosseini, H., & Conolly, M. J. (2013). A comprehensive survey of wearable and wireless ECG monitoring systems for older adults. *Medical & Biological Engineering & Computing, 51,* 485–495.

Cherkas, L. F., Harris, J. M., Levinson, E., Spector, T. D., & Prainsack, B. (2010). A survey of UK public interest in internet-based personal genome testing. *PLoS ONE, 5,* e13473.

Dennis, C. (2012). The rise of the narciss-ome. *Nature News.* https://doi.org/10.1038/nature.2012.10240.

Eggers, D. (2013). *The Circle.* London: Penguin.

European Science Foundation (ESF). (2012). *Personalised Medicine for the European Citizen—Towards More Precise Medicine for the Diagnosis, Treatment and Prevention of Disease.* Strasbourg: ESF.

Harris, A., Kelly, S. E., & Wyatt, S. (2016). *Cybergenetics. Health Genetics and the New Media.* London: Routledge.

Hiltzik, M. (2015). Need help paying for Amgen's Repatha? Get ready to give up your privacy. *Los Angeles Times*. http://www.latimes.com/business/hiltzik/la-fi-hiltzik-20151115-column.html. Accessed 27 July 2019.

Hodson, H. (2016). Revealed: Google AI has access to Huge Haul of NHS patient data. *New Scientist*. https://www.newscientist.com/article/2086454-revealed-google-ai-has-access-to-huge-haul-of-nhs-patient-data/. Accessed 27 July 2019.

Koenig, B. A. (2014). Have we asked too much of consent? *Hasting Center Report, 44*, 33–34.

Laurie, G. T. (2002). *Genetic Privacy: A Challenge to Medico-Legal Norms*. Cambridge: Cambridge University Press.

Matthews-King, A. (2015). GPs' diagnostic skills could be obsolete within 20 years, says Hunt. *Pulse*. http://www.pulsetoday.co.uk/political/political-news/gps-diagnostic-skills-could-be-obsolete-within-20-years-says-hunt/20030142.fullarticle. Accessed 27 July 2019.

Mayer-Schönberger, V., & Cukier, K. (2013). *Big Data*. London: John Murray Publishers.

McGoey, L. (2015). *No Such Thing as a Free Gift. The Gates Foundation and the Price of Philanthropy*. London: Verso.

McMahon, A., Buyx, A., & Prainsack, B. (2020). Big data governance needs more collective responsibility: The role of harm mitigation in the governance of data use in medicine and beyond. *Medical Law Review, 28*(1), 155–182.

National Academy of Sciences (NAS). (2011). *Toward Precision Medicine: Building a Knowledge Network for Biomedical Research and a New Taxonomy of Disease*. Washington, DC: NAS.

Neff, G. (2013). Why big data won't cure us. *Big Data, 1*, 117–123.

Office of the National Coordinator for Health Information Technology, The (ONC). (2015). *Connecting health and care for the nation: A shared nationwide interoperability roadmap*. https://www.healthit.gov/sites/default/files/hie-interoperability/nationwide-interoperability-roadmap-final-version-1.0.pdf. Accessed 27 July 2019.

Prainsack, B. (2015). Why we should stop talking about data 'sharing'. *DNAdigest Blog*. http://dnadigest.org/why-we-should-stop-talking-about-data-sharing/. Accessed 27 July 2019.

Prainsack, B. (2020). The value of healthcare data: To nudge, or not? *Policy Studies*. https://doi.org/10.1080/01442872.2020.1723517.

Prainsack, B., & Buyx, A. (2013). A solidarity-based approach to the governance of biobanks. *Medical Law Review, 21*, 71–91.

Prainsack, B., & Buyx, A. (2017). *Solidarity in Biomedicine and Beyond.* Cambridge: Cambridge University Press.

Rabinow, P. (1992). Artificiality and enlightenment: From sociobiology to biosociality. In J. Crary & S. Kwinter (Eds.), *Incorporations* (pp. 234–252). Manchester: Zone.

Spiegelhalter, D. (2013). Are you 45% more likely to die in a UK hospital rather than a US hospital? *British Medical Journal, 347,* f5775. https://doi.org/10.1136/bmj.f5775.

Taylor, M. (2013). *Genetic Data and the Law. A Critical Perspective on Privacy Protection.* Cambridge: Cambridge University Press.

The White House. (2015). *Factsheet: President Obama's precision medicine initiative.* Press Release. https://www.whitehouse.gov/the-press-office/2015/01/30/fact-sheet-president-obama-s-precision-medicine-initiative. Accessed 27 July 2019.

United Nations. (1966). *International convenant on economic, social and cultural rights.* http://www.ohchr.org/EN/ProfessionalInterest/Pages/CESCR.aspx. Accessed 10 May 2016.

Van Staa, T.-P., & Goldacre, B. (2016). Big health data: The need to earn public trust. *British Medical Journal, 354,* i3636.

Wachter, R. M. (2015). *The Digital Doctor: Hope, Hype, and Harm at the Dawn of Medicine's Computer Age.* New York: McGraw-Hill Education.

Weber, G. M., Mandl, K. D., & Kohane, I. S. (2014). Finding the missing link for big biomedical data. *The Journal of the American Medical Association, 33,* 2479–2480.

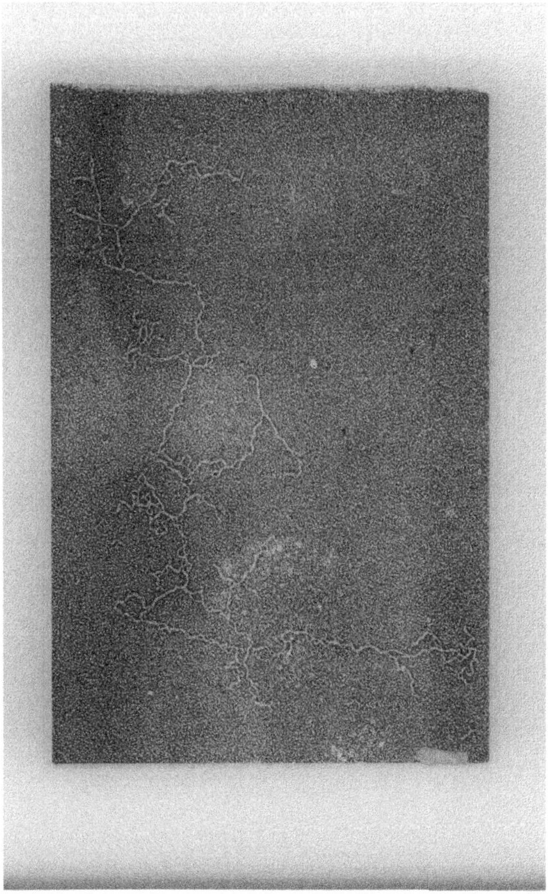

Photo 4 © Grischa Lichtenberger, 0815_11_sc_dia uni b_26

8

Experimenting with Solidarity in Biomedicine: From Practice to Principle?

Luca Chiapperino and Francesco Panese

In her contribution, Barbara Prainsack invites us to reflect on often-neglected facets of contemporary visions of personalised medicine, which emphasise less the power of genomics to give room to other, non-genetic information with potentially higher levels of actionability. From social media data, to wearable sensors, or even consumption patterns, the path towards data-intensive, individualised medicine is not only paved by the explanatory promises of molecular and systems biology, but draws heavily on categories of data not traditionally associated with medicine, which could yet prove fundamental in the development of monitoring

L. Chiapperino (✉)
Institute of Social Sciences—STSLab, University of Lausanne, Lausanne, Switzerland
e-mail: luca.chiapperino@unil.ch

F. Panese
Institute of Social Sciences—STSLab and Institute of Humanities in Medicine, University of Lausanne, Lausanne, Switzerland
e-mail: francesco.panese@unil.ch

D. Mahr and M. von Arx (eds.), *De-Sequencing*, Health, Technology and Society,
https://doi.org/10.1007/978-981-15-7728-4_8

and prevention strategies. As the author rightly points out—also through the mobilisation of the narrative scenario of the novel *The Circle*—the future depicted by using these data is no less controversial than the one in which genomics plays the role of the *soloist*.

In particular, two assumptions behind big data medicine catch the concerned reflection of Prainsack. On the one hand, what could be called the moral suasion of citizens with regard to their alleged duties to collect and share more data, according to which those who do not contribute their own data should actually be blamed for hampering the achievement of the societal good of better medicine. In her view, this line of reasoning is fundamentally flawed in that it puts on a par the responsibilities and accountability of those who would actually control data (e.g. biobanks, public agencies, hospitals, or even private companies), and those who instead could both benefit and be harmed by their use. A patient and a network of biobanks pooling data, a citizen and a company marketing genetic testing direct-to-consumer, Prainsack argues, are social actors whose duties and obligations with regard to the sharing and control of data are not the same. Thus any general call of duty to share and make accessible health data should take into consideration the imbalances of power currently existing in medicine, and devise effective collective strategies to protect the privacy of individual citizens. On the other hand, Prainsack's critique goes at the techno-epistemic core of the big data paradigm in biomedicine. The urge to share data in the discursive representations of personalised medicine often brushes over the concrete challenges of making data interoperable, and over the potential gains in terms of medically actionable information that could be obtained from data integration. Little is known, at this moment, about whether connecting heterogeneous datasets in medicine is an effort worth the huge investment it demands, or about the effective meaning and clinical benefits that could be obtained through this process.

Starting from this two-fold critique of personalised medicine, Prainsack argues for the need to innovate this paradigm democratically through social and political developments grounded in the normative framework of solidarity. In the understanding of its proponents (Prainsack and Buyx 2011), solidarity can be defined as the 'shared practices reflecting a collective commitment to carry "costs" (financial, social, emotional or otherwise) to assist others'. Solidaristic acts and practices are, in other words, not only individual expressions of sympathy towards others' vulnerability in which we recognise our own. Quite the contrary, solidarity often manifests itself in the form of solidified social norms and structures (e.g. welfare systems and social welfare arrangements) that represent an explicit societal commitment to assist others at a nominal cost for each member of a society. For this reason, this concept may prove—according to Prainsack—useful to account for the social and political challenges raised by personalised medicine. First, normative exercises of the solidarity principle could point to the need to bring the personal and collective levels together in the context of big data medicine, by pointing to the fact that self-interested and other-regarding actions are indeed not mutually exclusive in this context (data that are useful to one patient may also be useful for conditions afflicting another). Second, solidarity may highlight the multiple ways in which sharing one's data also constitutes a contribution to the achievement of collective goods. Third, solidarity-based approaches could also remind us that stratification should not lead to undue exclusion and should instead be driven by distributive considerations of benefit-sharing. Fourth, another important consideration stemming from the principle of solidarity is related to questions of access, and the imperative to ensure that people facing projects of big data medicine are not left alone should they decide not to, or be unable to, partake in these endeavours (Prainsack and Buyx 2017).

In our commentary, we focus on both the moral and the epistemic aspects of the critique presented by Prainsack, and complement her proposal with an emphasis on the relevance and value of practical experimentation with solidarity in personalised medicine that are both moral and techno-epistemic. As to the former, our objective is to complement her diagnosis on (what could be called) the moral economy (Fassin 2009) of personalised medicine with a reflection on the role of solidarity-based engagements and moral practices to counter the dystopia furthered through this paradigm. As to the latter, our goal is to show how working towards interoperability of data responds to distinct epistemic and techno-scientific objectives of biomedical actors, some of which can arguably be characterised as going in the direction of furthering solidarity-sensitive practices in our societies through personalised medicine.

8.1 Moral Economies Are Unstable Systems

In the book *At the Heart of the State*, Didier Fassin and his colleagues explore what they call the moral world of institutions (Fassin et al. 2015). Drawing from ethnographic studies of different institutional settings—from courts, the police, employment centres—the book provides us with vivid accounts of the lived experiences and moral functioning of various organizations in our societies. Following the work of practitioners in these different contexts allows them to reveal how institutions are 'the site where the state is produced' (p. 6); namely, how the 'macro-sociological and the micro-sociological levels' (p. 7) of state apparatuses are held together through the daily work, moral engagements, negotiations and relations of individual and collective agents working for them. Institutions are, in other words, the locus where the structure of a civic, legislative, and regulatory framework unfolds into and is (re)produced by habits, judgements, sentiments, punishments—in one word, morals—of individual and collective subjects. To capture these two intertwining levels of institution-making, the authors employ a distinction between moral economies and moral subjectivities. The former designate 'the production, circulation and appropriation of values and affects regarding a given social issue' in a specific polity, at a given historical moment (Fassin 2009; Fassin et al. 2015: 9). Roughly put, they constitute a common substantive understanding of the place, role, and evaluation of actors gravitating around a given social issue: they provide general institutional positioning, orient public sentiments and attitudes, and instruct individual and collective judgements of the situation. Moral subjectivities, by contrast, are the 'processes by which individuals develop ethical practices in their relationships with themselves or others' (Fassin et al. 2015: 9). They pertain to the level of the moral work involving individuals and collectives at the micro-level of routine activities in institutional settings, and typically consist of the moral tensions in the practice of turning the injunctions stemming from substantive representations of an issue (i.e. its moral economy) into embodied agency.

In this respect, moral subjectivities illustrate the lived body of a moral system: individual hesitations and pondering are the vivid solidification of the moral complexity of a situation, often neglected in its discursive and institutional representation. At the same time, however, moral subjectivities also reveal how agency is irreducible to the economy of sentiments, norms and principles giving meaning and value to a social phenomenon. The autonomy of agents to deviate from their professional ethos in order to lend a gesture of care, the decision to enforce more softly a given rule in the face of a particular situation, all testify to the generative power of subjective motivations, sentiments, and relations in the face of a structuring moral economy. In doing so, Fassin et al.'s book offers an in-depth demonstration of how moral economies are fundamentally unstable systems. While providing the substantive framing of a certain matter in our societies, their existence in the social space necessitates the composing, decomposing, and recomposing of their tenets by the concerned actors. This recognition, we argue, offers the opportunity for at least two reflections concerning Prainsack's socio-political uptake of the nascent paradigm of personalised medicine.

First, acknowledging the complex morality of institutional functioning suggests that—though the concerns expressed by Prainsack about personalised medicine are cogent—they belong to its moral economy, thus leaving open the possibility that relations, duties, care, and reciprocal obligations in given settings and situations may be otherwise. This is not meant to deny that the ideas and premises behind this data-intensive type of medicine can arguably be characterised as constituting a depersonalisation of medicine, reducing the individual patient to being a mere digital representation stripped of its complex social belonging and existential complexity (Zwart 2016). Nor we mean to mitigate the parallel critique showing how personalisation, as coveted medical reality, often goes hand in hand with a problematic rhetoric of empowerment conceiving a devolution of responsibility to the citizenry with regard to their health (Juengst et al. 2012; cf. Prainsack 2017). Nor, finally, it is an exercise of wishful thinking denying the importance of hypes, programmatic visions, soft regulations, policy visions, and biomedical in the practical unfolding of innovation (Hedgecoe 2004). Rather, what we mean here is to highlight the gap between the general normative undertones of claims supporting the vision of personalised medicine and the complexities faced by the subjectivities working on the transformation of personalisation into a reality of medicine. If we hold this distinction, we can see that the future of personalised medicine is not given, that the problematic social circulation of particular technologies of monitoring and data brokering is not inevitable, that the moral fabric of personalised medicine is, in large part, still to be made. By holding a methodological distinction between moral economies and subjects, we do not have to accept the dystopian version of personalised medicine criticised by Prainsack, mostly because the realisation of this paradigm is still open, still undecided: too many competing values, sentiments, desires, expectations (are likely to) play a fundamental role in the practical configurations this vision of medicine is going to engender.

Against this backdrop, Prainsack's work constitutes an important starting point. Her call for solidarity-based arrangements that could innovate democratically the domain of healthcare has the merit of highlighting alternative views of the moral economy of personalised medicine, which could instruct regulations, policy frameworks and moral orientations genuinely serving the interests of public health. Yet, such solidarity-based thinking calls for a scrutiny of the subjectivities currently experimenting with the moral economy of personalised medicine. As we have seen in the example of Fassin and his colleagues, the moral work of citizens, patients, practitioners and experts involved in personalised medicine-related initiatives goes beyond the simple reproduction of the moral suasions and concealed disempowerment detectable at a discursive level. The *moral world of personalised medicine*—if we wanted to paraphrase Fassin and colleagues —may rather be defined by the substantive tension between such generative practices and structuring discourses.

This leads us to a second reflection, which pertains to the strategies to further Prainsack's normative proposal in the nascent socio-political configurations of future medicine. What role could solidaristic practices, that may be currently part of the daily, mundane work of experimenting with the paradigm of personalised medicine, play in the promotion of her solidarity-based moral economy of future healthcare? As shown by

Hoeyer et al. (2017) in the context of an international collaborative project, the material and informational flow of biomedical research often demands a type of moral engagement on the part of researchers that goes beyond the principles and regulations that are the norm in their professional practice. While the latter, as Hoeyer and colleagues show, are very useful to inform researchers about what they are allowed to do, it is rather through the 'morality built into ties between people' (Hoeyer et al. 2017: 393) that research collaborations among scientists and donors continue to flow. This recognition goes in the direction of the distinction we have drawn from Fassin and colleagues between moral economies and subjectivities, and raises the question as to whether the same may currently be happening with regard to solidarity in personalised biomedicine. Regardless of whether this principle is integral or foreign to the moral economy of personalised medicine in a given socio-political setting, solidarity could be seen—in this vein—as one among other values that are being negotiated in the daily subjective work of researchers experimenting with personalisation. And, indeed, some evidence to support this view may be found in the literature. From the decision to implement institutional policies for data restitution in genomic research that take into consideration vulnerable (biological) relatives of sample donors, while not required by current legislation (Hens et al. 2011), to the development of alternative models of informed consent that deviate from the gold standard of general consent to give more room to patient participation, and partnerships between them and the researchers (Blasimme and Vayena 2016), several elements suggest that solidaristic practices are already part and parcel of the moral world of personalised medicine. This posits the opportunity of a thoroughly complementary strategy for the promotion of Prainsack's proposal, and one that starts methodologically from a problematisation of the hiatus between the complexity of moral practices and the economies of values that guide biomedicine. Ultimately, the importance of promoting solidarity-based arrangements in personalised medicine is a strong political argument for Prainsack, which—at the same time—may be a very practical concern for researchers and practitioners currently working on the translation of personalisation into a medical reality. What if, then, we needed to generalise more moral subjective experimentations with solidarity than to promote it as a principle in personalised medicine? To paraphrase Bernard Williams' critique of Kantian moral philosophy, future patients may need the benefits of gestures and practices of solidarity, even more than the benefits of its principles (Williams 1999).

8.2 The Epistemic Work Towards Interoperability and Solidarity

The same kind of reasoning could apply to the epistemic and techno-scientific fabric of personalised medicine, and consequently to the critique Prainsack moves to the epistemic value of interoperability in personalised medicine. The visions of data-intensive individualised medicine she discusses are in fact currently being translated into concrete techno-scientific assemblages that are pivotal to the establishment of personalised medicine as translational research and biomedical practice in given socio-political settings. In the Swiss case (Meier-Abt et al. 2015), for example, it is also possible to underline a major discursive investment into interoperable data infrastructures (Swiss Personalised Health Network 2017). Building an infrastructure of interoperable datasets is in fact the main declared objective of Swiss initiatives, which is articulated into different concrete initiatives, including 'the molecular analysis of samples, the sharing of data for a large population, the development of standardised bioinformatics tools … and finally, the interpretation of these data enabling optimal therapeutic decision-making' (Coukos et al. 2014: 1099). A closer look at these different research, translational, and clinical endeavours reveals how interoperable data infrastructures—more than constituting a single object of research and practice—constitute a distant objective (cf. Guchet 2016) towards which these actors work by negotiating their experimental, technological, social, economic, legal, and political facets. A few examples may be useful here to illustrate how interoperability of data in the infrastructures at the centre of this and similar endeavours constitute concrete elements of professional work towards the alignment of various techno-scientific promises with medical reality.

Driven by the imperatives of research discoveries and the development of innovative therapeutics or biomarkers, those who have developed platforms of biobanking for sample storage and sharing (Mooser and Currat 2014) consider the integration of data as a fundamental resource to harness the potentialities of current biomedical technologies, and to streamline the research process of development for new preventive and therapeutic strategies. Without going as far as considering it able to yield meaningful, medically actionable information—something Prainsack arguably underlines as being unwarranted – practising interoperability is in this context building an environment favourable to research towards personalised medicine. Interoperability means here establishing conventions of practices and norms (e.g. common ontology of different data, homogeneous consent standards, benchmarking of sample processing and data interpretation, etc.) among citizen-donors and the community of clinicians and researchers. Differently from this example, interoperability is instead a matter of hermeneutics and interpretation for those working toward the development of bioinformatics tools for data analysis (Lescuyer et al. 2007; Sansone et al. 2012). Rather than pointing to the relational function of interoperable biological data, in this concrete engagement what is at stake is data integration, that is, the combination of different layers of biological data in a meaningful way. Bioinformatic interoperability of multi-omic data—but also digital health data collected through technologies such as trackers, sensors, or wearables (see Salathé et al. 2012; Lovis and James 2016; Trono 2016)—finds its rationale in the capacity to describe interaction between phenomena happening at different level of biological organization. In this respect, the promises attached to the operationalisation of interoperability seem to sketch a dynamic that is symmetrically different from the one we encountered in the example above. While in the first case, interoperability operates as an invisible, background feature of the research ecology pivoting around biobanking, here interoperable datasets are rather supposed to resurface systematically—or even automatically, if we take at face value the promissory rhetoric attached to artificial intelligence (Howard and Huber 2016)—as an informed grasp of the relevant property of biological systems and their variation from a statistical average.

Needless to say, the question of medical actionability of this information emphasised by Prainsack is yet another layer of complexity in the different pragmatic engagements with data interoperability we have illustrated briefly in the Swiss context. However, what the examples above show is that building infrastructures for data-driven medicine is a layered, complex and heterogeneous practice, informed by distinct epistemic and techno-scientific ends, as well as visions as to why it is worth producing, pooling, and implementing knowledge for the personalisation of future medicine. A process that, as we have seen, gives shape to distinct pragmatic functions of interoperability, and that, we might add, is also likely to be thoroughly imbued with the subjective experimentations with the moral economy of personalised medicine highlighted in the previous section. This is the next (and last) piece of the jigsaw portrayed by Prainsack upon which we wish to elaborate.

The epistemic engagements with interoperability and sharing of data can indeed also be characterised as a plurality of normative and value-laden engagements with the socio-political configurations of personalised medicine. Building an infrastructure of interoperable data is, in this vein, not just creating an ecosystem allowing researchers, patients, and clinicians to pursue different epistemic objectives, but also an exercise of co-construction of the *moral world* of such infrastructures. One example—always drawn from our observations in the Swiss context—can illustrate these moments of co-production (Jasanoff 2004) in personalised medicine: the attempt to provide technical solutions to the ethical issues of privacy and re-identification in IT infrastructures of personalised data sharing. Current configurations of research, and the public debate ('Génome chiffré, génome protégé' 2017) around Swiss initiatives towards personalised health revolve partly around the relevance and utility of strategies for the encryption of the data collected. Developing effective cryptography standards is, in this context, not just an engineering problem for the construction of informatics systems of data-sharing (Check Hayden 2015), but rather constitutes the opportunity to infuse the epistemic machinery of personalised medicine with values of digital trust and privacy-conscious technological standards (Barraud 2017). Striking the right balance between circulation of different data, minimisation of risks for re-identification, and prevention of breaches of access through a technical standard for data handling and sharing is, in other words, a way to promote trusted participation in this endeavour, and to prevent indiscreet access to these data by unwarranted actors. Specifically, the interests of insurance companies and the reported cyber attacks to which healthcare institutions are subjected in the US ('Health Care Industry Cybersecurity Task Force—PHE' 2018) provide, in the case at issue, a moral justification for developing robust encryption standards in interoperable data systems. In brief, pursuing the techno-epistemic objective of a secure and interoperable data infrastructure can reasonably be characterised as one facet of the moral work towards an equitable and privacy-promoting configuration of personalised medicine.

The brief example of encryption shows how developing interoperable data infrastructures also means aligning techno-epistemic standards with values of trust in the healthcare system and with the political struggles to prevent potential discrimination against patients and citizens. By ensuring that any illegitimate user of health data is kept out of the infrastructure, standards of encryption represent a resource for biomedical actors to sketch a specific social circulation of their work towards personalised medicine. And one, we argue, that could go in the direction of rendering solidaristic practices *both a moral and a techno-scientific condition of possibility* of personalised medicine. The normative and technical imperative of cryptography could in fact arguably be seen as no less than one—limited, yet significant—strategy to ensure that openness of health data in the future of medicine does not undermine existing solidaristic structures in healthcare. In this respect, Prainsack's critique of interoperability and data sharing in these endeavours is an important reminder of how 'a critical examination of the assumptions and consequences of such an approach in medicine is an important task'. Yet, what our analysis has tried to show is that interoperability may not only be valuable to a pre-given and static rhetoric worth of Prainsack's critiques, but also may qualify as a symmetric and complementary practical engagement with the moral filigree of data-intensive and personalised medicine. Weaving the socio-technical fabric of personalisation in medicine does not rest solely on the problematic economy of values, norms and sentiments criticised by Prainsack. Rather, it is a practical process straddling the social and techno-epistemic dimensions of this endeavour, and one that—in at least one telling example—displays material and technological experimentations that could be compatible with her call to promote solidarity in the social structure and power dynamics of medicine.

8.3 Conclusions

The contribution by Barbara Prainsack invites us to reflect on the socio-political innovations that the normative exercises of solidarity could bring to the realisation of personalised medicine. In the present commentary, we have highlighted multiple ways in which solidaristic practices could be observed in the unfolding of personalisation in medicine at both the moral and epistemic levels. In this respect, our analysis can be regarded as a complement to Prainsack's critique of the moral economy of data-intensive and personalised medicine. Specifically, our work constitutes an invitation to consider the framework proposed by the author, not just as a positive normative guidance in the societal appraisal of socio-political initiatives towards personalised medicine. Rather, our focus on micro-level experimentations with personalisation reveals how solidarity in thinking could also qualify as a productive methodological operator inviting us to reconstruct how, amid the yet indeterminate configurations of future medicine, its value is being negotiated in, and thus could flourish from the praxis of making personalised medicine a social reality.

References

Barraud, E. (2017). *EPFL teams up with partner organizations to build 'digital trust'*. https://actu.epfl.ch/news/epfl-teams-up-with-partner-organizations-to-build-/. Accessed 27 July 2019.

Blasimme, A., & Vayena, E. (2016). Becoming partners, retaining autonomy: Ethical considerations on the development of precision medicine. *BMC Medical Ethics, 17,* 67. https://doi.org/10.1186/s12910-016-0149-6.

Check Hayden, E. (2015). Extreme cryptography paves way to personalised medicine. *Nature News, 519,* 400. https://doi.org/10.1038/519400a.

Coukos, G., Michielin, O., Dietrich, P.-Y., & Aapro, M. (2014). Ensemble vers L'oncologie de précision!'. *Revue Médicale Suisse, 431*, 1099–1100.

Fassin, D. (2009). Moral economies revisited. *Annales. Histoire, Sciences Sociales, 64*, 1237–1266.

Fassin, D., et al. (2015). *At the Heart of the State: The Moral World of Institutions*. London: Pluto.

'Génome chiffré, génome protégé' 2017: 24. https://www.letemps.ch/sciences/genome-chiffre-genome-protege. Accessed 28 July 2019.

Guchet, X. (2016). *La Médecine Personnalisée. Un Essaie Philosophique. Médecine et Sciences Humaines*. Les Belles Lettres. https://www.lesbelleslettres.com/livre/1277-la-medecine-personnalisee. Accessed 28 July 2019.

Health Care Industry Cybersecurity Task Force—PHE. (2018). https://www.phe.gov/preparedness/planning/CyberTF/Pages/default.aspx. Accessed 28 July 2019.

Hedgecoe, A. (2004). *The Politics of Personalised Medicine: Pharmacogenetics in the Clinic*. New York: Cambridge University Press.

Hens, K., Nys, H., Cassiman, J. J., & Dierickx, K. (2011). The return of individual research findings in paediatric genetic research. *Journal of Medical Ethics, 37*, 179–183. https://doi.org/10.1136/jme.2010.037473.

Hoeyer, K., Tupasela, A., & Bøgehus Rasmussen, M. (2017). Ethics policies and ethics work in cross-national genetic research and data sharing: Flows, nonflows, and overflows. *Science, Technology and Human Values, 42*, 381–404. https://doi.org/10.1177/0162243916674321.

Howard, P., & Huber, P. (2016). *Unlocking Precision Medicine*. New York: Encounter Books.

Jasanoff, S. (2004). *States of Knowledge: The Co-production of Science and the Social Order*. London: Routledge.

Juengst, E. T., Flatt, M. A., & Settersten Jr., R. A. (2012) Personalised genomic medicine and the rhetoric of empowerment. *The Hastings Center Report, 42*, 34–40. https://doi.org/10.1002/hast.65.

Lescuyer, P., Hochstrasser, D., & Rabilloud, T. (2007). How shall we use the proteomics toolbox for biomarker discovery? *Journal of Proteome Research, 6*, 3371–3376. https://doi.org/10.1021/pr0702060.

Lovis, C., & James, C. (2016). Santé Digitale: Petit Guide À L'usage Du Néophyte. *Revue Médicale Suisse, 12*, 1108–1112.

Meier-Abt, P. J., SIB, R. A., Frey, U., Günther, D., Mooser, V., Probst-Hensch, N., et al. (2015). *Implementation of the Swiss Personalised Health Network (SPHN) initiative*. https://www.sbfi.admin.ch/dam/sbfi/de/dokumente/report_implementationoftheswisspersonalisedhealthnetworksphninit.pdf.dow nload.pdf/report_implementationoftheswisspersonalisedhealthnetworksphninit.pdf. Accessed 28 July 2019.

Mooser, V., & Currat, C. (2014). The Lausanne Institutional Biobank: A new resource to catalyse research in personalised medicine and pharmaceutical sciences. *Swiss Medical Weekly*. https://doi.org/10.4414/smw.2014.14033.

Prainsack, B. (2017). *Personalised Medicine: Empowered Patients in the 21st Century?*. New York: New York University Press.

Prainsack, B., & Buyx, A. (2011). *Solidarity: Reflections on an Emerging Concept in Bioethics*. Citeseer.

Prainsack, B., & Buyx, A. (2017). Solidarity in practice II: Personalised medicine and healthcare. In *Solidarity in Biomedicine and Beyond* (pp. 123–144). Cambridge: Cambridge University Press. http://ebooks.cambridge.org/ref/id/CBO9781139696593. Accessed 29 July 2019.

Salathe, M., Bengtsson, L., Bodnar, T. J., Brewer, D. D., Brownstein, J. S., Buckee, C., et al. (2012). Digital epidemiology. *PLOS Computational Biology, 8*, e1002616. https://doi.org/10.1371/journal.pcbi.1002616.

Sansone, S., et al. (2012). Toward interoperable bioscience data. *Nature Genetics, 44*, 121–126. https://doi.org/10.1038/ng.1054.

Swiss Personalised Health Network. (2017). *Factsheet SPHN at a glance*. http://www.samw.ch/dam/jcr:962d21ee-870e-4989-b81d-71605b454980/factsheet_sams_sphn_2017.pdf. Accessed 28 July 2019.

Trono, D. (2016). Switzerland and the digital health revolution. *CHIMIA International Journal for Chemistry, 70*, 851–852. https://doi.org/10.2533/chimia.2016.851.

Williams, B. A. O. (1999). *Problems of the Self: Philosophical Papers 1956–1972*. Cambridge: Cambridge University Press. Reprint.

Zwart, H. A. E. (2016). The obliteration of life: Depersonalization and disembodiment in the terabyte era. *New Genetics and Society, 35*, 69–89. https://doi.org/10.1080/14636778.2016.1143770.

9

The Moral Making of Data-Rich Personalised Medicine

Gaia Barazzetti

Building on the futuristic scenarios presented in the dystopian novel *The Circle*, Barbara Prainsack invites us to reflect on an ongoing transformation in the way we collect and use health-related information for research and medical purposes, which is potentially disruptive to our conceptions of health and disease as well as to our health systems. Such a transformation goes hand in hand with current data-rich representations of *precision* medicine which, as Prainsack tells us, promises to move away from the old symptoms-based taxonomies of disease and the *one-size-fits-all* approach to healthcare, and to tailor preventive and therapeutic interventions to the individual characteristics of each patient. Resting on the potential of big data analytics, this new approach to medicine and healthcare envisages going far beyond the collection of biological samples or genomic information, to include the ever-increasing amount of health-related data being produced, not only in healthcare but also by individuals themselves in their everyday social media, health and fitness

G. Barazzetti (✉)
The ColLaboratory, University of Lausanne, Lausanne, Switzerland
e-mail: gaia.barazzetti@unil.ch

D. Mahr and M. von Arx (eds.), *De-Sequencing*, Health, Technology and Society,
https://doi.org/10.1007/978-981-15-7728-4_9

self-tracking, or consumer activities, captured on the internet or through connected objects. Advocates of data-rich precision medicine claim that the data will help to improve diagnosis and treatment significantly, to bridge biomedical research and healthcare successfully, and to produce predictive knowledge that will allow a shift from disease and treatment to health and prevention before any symptoms might appear.

As Prainsack tells us, health data mining is presented as *ethically* trivial: there is so much data out there that can potentially be useful in improving individual and public health, and can obviously be worth collecting and processing for research purposes. The arguments put forward by proponents of precision medicine is even more radical, as they claim that we have a moral duty, as individuals and as a society, to collect and use health data in research, and that it would be unethical not to do so. In the face of discourses aimed at exalting the ethical evidence of mining and sharing individual's health data, Prainsack makes some important arguments to unpack two basic assumptions in current visions of data-rich precision medicine: the idea that there is a moral duty to collect and share data, and the belief that genomic and health-related data can easily be made interoperable and meaningful in healthcare.

In my commentary, I shall focus on one key element put forward in Prainsack's arguments: the deployment of a moral vocabulary in current visions of data-rich precision medicine. Appeals to ethical values are so ubiquitous in this field that they seem to be part of what can be called a moral *economy* (Fassin 2009; Fassin and Eidelman 2012), which plays a fundamental role in building the credibility of this new approach to health and disease in the eyes of policy-makers, healthcare professionals and the public.

9.1 Personalising Medicine for the 'Common Good'?

As Prainsack shows, the supposed ethical obligation to foster the mining and sharing of genomic and other health data is based on the idea that precision medicine will dramatically increase both individual and public health benefits. As data-rich precision medicine does not rest solely on the discovery of new personalised drugs and diagnostic tools, but aims to stratify treatments and screening programmes based on those who may actually benefit from them. It is assumed that this will help to reduce current healthcare costs by improving the efficacy and mitigating the potential harm of medications and screening procedures. The rhetoric associated with the *common*—individual as well as collective—good that is expected to be so powerful that public health benefits will override those of individuals, and become immediately evident as being valuable.

Following Prainsack in her critical examination of the moral assumptions behind sharing data for research, I would like to pinpoint some critical aspects in the notion of the common good that is put forward in data-rich representations of precision medicine. This idea of the common good at stake is based on the assumption that *knowledge* that can be gathered from the analysis of genomic and other health related data is clinically meaningful and valuable, and that, as a consequence, this knowledge goes together with the *power* to act, through treatment or prevention measures, and with a *moral duty* to do so. This knowledge, however, needs to demonstrate its clinical *validity* and *utility* (Foster et al. 2009; Edelman and Eng 2009; IOM 2012) in order to translate into concrete healthcare benefits for the individual and for society. The clinical validity of such information essentially resides in its predictive value, or rather in the probability of the correlation between, e.g., a positive result from a disease susceptibility test and the onset of that illness. To date, one of the greatest difficulties linked to the evaluation of clinical validity concerns the interpretation of predictive information on most common diseases, which are multifactorial as they depend on the complex relationship of genetic and environmental factors (Shirts and Parker 2008). In such cases, the interpretation of the information can vary considerably over time with the advancement in knowledge about genomics and health risks. As a consequence, health professionals should not only have the updated knowledge necessary to assess the clinical validity of the information, but should also appeal regularly to more specialised competences in genomics and big data health analytics to verify and update data interpretations. Similarly, health professional organisations and public health institutions responsible for the training and practice of health professionals should, for their part, adapt training programmes and elaborate guidelines for the evaluation and periodic revision of the predictive value of information gathered from data analysis. All this endeavour would require, obviously, that human and economic resources necessary to such changes in healthcare training and practice are available, which is not absolutely evident for the great majority of health care systems today.

With regard to the clinical *utility* of the knowledge that can be grasped from the analysis of genomic and other health data, its assessment should ideally include, in addition to the reduction in morbidity and mortality, also its contribution to the well-being of the individual. In this sense, the evaluation of the clinical utility should involve an appreciation of the *personal* utility (Bookman et al. 2006; Ravitsky and Wilfond 2006; Bunnik et al. 2014; NAS 2018), for the individual, of the information on, e.g., disease susceptibility. However, studies conducted in genetic medicine have already shown how knowing about genetic health risks does not necessarily lead to patients changing their habits and adopting healthier life styles (Henrikson et al. 2009). In particular, the practice of genetic counselling has shown that the relationship between knowledge about genetic risks and individual well-being can be complex and difficult to elaborate for the patient (Clarke and Thirlaway 2011).

Discussion about coping with genetic risks highlights a dichotomy between two approaches to *personalisation*, which have been defined in terms of *personal-molecular* and *personal-subjective*. While a *personal-subjective* perspective acknowledges the boundless uniqueness of every individual and is more centred on a holistic approach to patient care, taking into account the psychosocial and spiritual aspects of individual well-being, a *personal-molecular* approach to personalisation consists in focusing on the molecular characteristics of the individual and fragmenting the person into a plurality of objectifiable, measurable and actionable biomolecular benchmarks. As shown by philosopher Xavier Guchet (2016), the polysemy associated with the notion of personalisation risks to produce misconceptions and false expectations among patients who, confronted with information about health risks, feel directly concerned and have to make decisions for themselves. From the point of view of the patient, in fact, this information does not make sense when compared with the massive amount of aggregated data used to extrapolate predictions in terms of disease susceptibility and health optimisation. Rather, this information is meaningful to patients because of its implications for their own lives and interpersonal relationships.

Thus the notion of common good associated with visions of data-rich precision medicine can be questioned not only in terms of public health benefits, but also with regard to the personal benefit for the patient. In this sense, Prainsack's claim to stop being blinded by the moral rhetoric of sharing and instead clarify the conditions and the purposes for massively collecting and sharing health data invites us to engage, as a society, in a public debate about the common good we would like to pursue with our well-meant sharing efforts.

9.2 Individual Responsibility for Our Collective Medical Future

Prainsack's criticism of the moral assumptions behind current representations of precision medicine clearly show how visions of personalised medical futures goes hand in hand with a rhetoric of patient's empowerment that is presented as good and evident in itself, as it is opposed to medical paternalism and based on the idea that the basic availability of information about health risks corresponds to the capacity of anyone to prevent disease and stay healthy. This vision, however, accentuates individual responsibility at the expense of a consideration of the socio-economic factors that can determine the capability of the person to endorse such a responsibility towards his/her own health. In this sense, data-rich personalised medicine fits into what sociologist Nikolas Rose (2007) has defined as a more general transition in contemporary biomedicine, from the management of health risks at the population level, to the individualisation of risks, through an ever increasing responsibilisation of individuals. The value granted to individual responsibility in public health management is not recent and traces back to increasing criticisms, starting from the later decades of the nineteenth century, towards European welfare politics, which were considered inadequate to contain an increase in public health expenditure and were supposed to favour a generalised de-responsibilisation as they might exacerbate individuals' dependence on the support of the state. Welfare reforms have therefore emphasised the promotion of autonomy and empowerment of individuals, based on the idea that it would be preferable, both economically and ethically, to place more responsibility on the individual. Appeals to the individuals' empowerment and responsibility for their health, however, risk turning into a *double edged sword* (Rose 2013): on the one hand, the availability of information on health risks and disease susceptibility may increase people's capacity to endorse better lifestyle behaviours; but on the other hand, everyone would be expected to conform to social expectations regarding the use of this information to prevent disease and take care of one's own health as far as possible. Rose paints a rather dark scenario, where individual responsibility for health would reach the extreme forms of social pressure to continuously self-monitor health and life style, and where individuals would be blamed, or even considered as guilty, for failing to translate their knowledge of health risks into a *power* and a *duty* to stay healthy.

Taking us far beyond Rose's scenario, and in much closer contact with the real world of powerful corporations dominating health data-mining globally, Prainsack highlights how moral appeals to responsibilisation have extended from individual responsibility to a societal responsibility for the health of the population, with national governments reproducing the moral rhetoric of data sharing without calling into question the power gap between public and private actors. Prainsack shows us very vividly how state actors' subjection to the economic power of GAFAM (i.e. Google, Amazon, Facebook, Apple, Microsoft) contributes to seal the inevitability of this transformation in the management of individual and public health. Prainsack reminds us that public actors should instead assume their political responsibility with respect to these drifts, precisely because, if individual empowerment and participation are important, the role of the state is to protect and promote individual rights and the common good with respect to powerful economic actors who could prevail over them. 'The way out of this is through us,' she also reminds us, by calling on us to take on the responsibility that each of us has, as a citizen, to engage in a democratic debate on the common good at stake, and on the values we would like to guide our collective medical future.

References

Bookman, E. B., Langehorne, A. A., Eckfeldt, J. H., Glass, K. C., Jarvik, G. P., Klag, M., et al. (2006). Reporting genetic results in research studies: Summary and recommendations of an NHLBI working group. *American Journal of Medical Genetics, 140A*, 1033–1040.

Bunnik, E. M., Janssens, A. C., & Schermer, M. (2014). Personal utility in genomic testing: Is there such a thing? *Journal of Medical Ethics, 41*, 322–326.

Clarke, A., & Thirlaway, K. (2011). Genetic counselling for personalised medicine. *Human Genetics, 130*, 27–31.

Edelman, E., & Eng, C. (2009). A practical guide to interpretation and clinical application of personal genomic screening. *British Medical Journal, 339*, b4253.

Fassin, D. (2009). Les économies morales revisitées. *Annales, Histoire, Sciences Sociales, 6*, 1237–1266.

Fassin, D., & Eidelman, J.-S. (2012). Défense et illustrations des économies morales. In D. Fassin, et al. (Eds.), *Economies morales contemporaines* (pp. 9–18). Paris: La Découverte.

Foster, M. W., Mulvihill, J. J., & Sharp, R. R. (2009). Evaluating the utility of personal genomic information. *Genetics in Medicine, 11*, 570–574.

Guchet, X. (2016). *La médecine personnalisée. Un essai philosophique*. Paris: Les Belles Lettres.

Henrikson, N. B., Bowen, D., & Burke, W. (2009). Does genomic risk information motivate people to change their behavior? *Genome Medicine, 1*, 37.

IOM (Institute of Medicine). (2012). *Evolution of Translational Omics: Lessons Learned and the Path Forward*. Washington, DC: The National Academies Press.

NAS (National Academies of Sciences, Engineering, and Medicine). (2018). *Returning Individual Research Results to Participants: Guidance for a New Research Paradigm*. Washington, DC: The National Academies Press.

Ravitsky, V., & Wilfond, B. S. (2006). Disclosing individual genetic results to research participants. *American Journal of Bioethics, 6*, 8–17.

Rose, N. (2007). *The Politics of Life Itself: Biomedicine, Power, and Subjectivity in the Twenty-First Century*. Princeton, NJ: Princeton University Press.

Rose, N. (2013). Personalised medicine: Promises, problems and perils of a new paradigm for healthcare. *Procedia—Social and Behavioral Sciences, 77*, 341–352.

Shirts, B. H., & Parker, L. S. (2008). Changing interpretations, stable genes: Responsibilities of patients, professionals, and policy makers in the clinical interpretation of complex genomic information. *Genetics in Medicine, 10,* 778–783.

Photo 5 © Grischa Lichtenberger, 0815_11_sc_dia uni b_6

Part IV

Families

10

An Ordering of Letters: My Own Personal Genome

Kate O'Riordan

The language of personal genomics promises us all our own personal genome. One of the contradictions of this promise is that it operates as though we did not already embody its materials ourselves. This promise, then, is one of a gift to peoples, created and curated for us by direct-to-consumer genomics companies, or national genome projects. This contradiction, in promising to give us something we already have, but cannot access without biotechnology, brings into focus the mediation of the genome: its representational, communicative, and technological in form and content. The promise of the genome, as opposed to the gene, is that it is informational, liberating, and democratic (Reardon 2012). It is also indefinitely meaningful, promising to improve the future materially while remaining symbolic in the present. Despite the high-tech imaginary of genomics, however, the language of promise in relation to genomics is ordered through very familiar language. Today, for example,

K. O'Riordan (✉)
School of Media, Arts and Humanities, University of Sussex,
Brighton, UK
e-mail: k.oriordan@sussex.ac.uk

D. Mahr and M. von Arx (eds.), *De-Sequencing*, Health, Technology and Society,
https://doi.org/10.1007/978-981-15-7728-4_10

headlines read that genome editing breakthroughs are the holy grail of genetics (Johnston 2016). In 1991, some thirty years ago, the holy grail of genetics was the Human Genome Project (Nuttall 1991), a project that sees us as genomic rather than genetic. However, the ordering of letters persists. The ordering of language constructing master genes, genetic determinism, broken genes, genetic mistakes, and holy grails is still attached to the letters of the nucleotides of DNA. The ordering of letters remains part of the choreography of the address, from the language used to communicate genomics to the public, to the letters of the genome, to the letters of communication between healthcare professionals and patients.

Human genomics is communicative and has involved a proliferation in life stories (O'Riordan 2011, 2016). It has been taken up in the making, and unmaking, of stories about identity, ancestry, biography, and autobiography. Genome sequencing read in this light can be seen as a technology of life story writing. Popular television genealogy programming has taken up DNA testing enthusiastically, and direct-to-consumer testing and sequencing has featured as part of celebrity life story across news, documentary, books, and television forms. Stories about genomic identity narrate stories of origin, reinforcing or disrupting other forms of story. They also project mortality and modes of dying.

One of the challenges of the life story as a mode of telling tales is that the narrative of the individual protagonist is inevitably a form of individualisation or exceptionalism. However, individual stories also offer points of identification—connection as well as difference. Genomics invites intimacy at the same time as it is structural. Like identity itself, it is both singular and collective, reinforcing categorisations such as race, ethnicity, and class, while at the same time offering the possibility that these might be undone. Looking at multiple stories, and different scales of story, helps to bring into focus the structural effects of inequality, privilege, and precarity. Sometimes family and patient group stories can do that work, and sometimes they also reinforce individual exceptionalism. This essay is an attempt to take up threads of story, of life story, and entanglement to open up some of the detail of experience and to put this into conversation with the promise of personal genomes.

10.1 100,000 Personal Genomes

Genetic testing has become overshadowed by genome sequencing in biomedical research. The Human Genome Project of the 1990s increased the range of tests available. Sequencing involves magnitudes of data and sequencing projects seek thousands of genomes rather than individual ones. Genetic testing and sequencing after the Human Genome Project draws people into biomedical projects as research subjects, under the banner of health benefits and economic promise for nation states and shareholders. Personal genomics, self-tracking, and genetic testing all invite responsibility for health care and DIY health promotion, and disease management and prevention. The personal, as in personal genomics, is invoked at the same time as 100,000 genomes becomes the relevant data set for many contemporary genome sequencing projects. In the UK, for example, the national flagship sequencing project Genomics England promises to sequence 100,000 genomes.

Genome sequencing then operates at the personal scale of life story while at the same time on the industrial scale of big data, in the order of petabytes (10^{15}), next-generation sequencing, and server farms. This story is an attempt to scale up and down, to trace the touch of genomics in everyday life, combining biography and autobiography to tell a tale of a collective experience in which the industrial scale of personal genomics seemed to foreclose on a life, and undo subjects.

10.2 Life Stories

In 2009, I began sharing a house with my mother, for the first time in 20 years. We rented a large town house that could accommodate three generations without anyone feeling too crowded. My mother, who lived at the top of the house, referred to herself as the granny in the attic. Soon after moving in, though, mum had issues with balance and vertigo, coughing and generally not feeling great. Maybe the attic and the very steep stairs were bad ideas. More than one person broke a foot or toe during that time. Mum wasn't very keen to seek medical advice but after some persuasion she did. After a few consultations with the doctor at her local surgery, the consensus was that she should undergo a range of investigatory tests to try to figure out what was going on. As a result of this, mum went into hospital for a couple of days to get everything done at once. Many tests later she came home.

A few weeks after this experience she received a letter with an appointment for meeting with a genetic consultant. The appointment had been arranged on the basis that the tests had revealed the presence of a genetic condition: Spinocerebellar ataxia type 6 (SCA6). The link between the gene CACNA1A and SCA6 was established during the Human Genome Project (Zhuchenko et al. 1997). SCA6 is caused by CAG repeats in CACNA1A, where normal alleles have 4–18 repeats and pathogenic alleles are thought to have 19–33 repeats. The number of repeats is also thought to give an indication of onset (Fig. 10.1).

This chapter details experiences around this testing, the decisions made by different people in the family, and the experience of the condition itself. It also explores the consequences of the enthusiasm of the genetic consultant about the condition, and subsequent overlooking of other evidence. This enthusiasm led to a more urgent and acute condition being missed. This story then becomes one of different diagnoses and tests and their reception, as well as a story about the final part of my mother's life before she died of the effects of lung cancer, and its treatment.

On the one hand this is a story of a number of mistakes, some of which were errors in judgement. It is also, on the other, the story of how a person's life expectancy suffered adversely in direct relation to the promise of genomic futures. It is an example of where the value of high-tech novel tests and recent discoveries were valued over more basic

SPINOCEREBELLAR ATAXIA 6; SCA6

Phenotype-Gene Relationships

Location	Phenotype	Phenotype MIM number	Inheritance (in progress)	Phenotype mapping key	Gene/Locus	Gene/Locus MIM number
19p13.13	Spinocerebellar ataxia 6	183086	AD	3	CACNA1A	601011

Fig. 10.1 SCA6 entry in the OMIN database

measures (X-ray) and it is about vulnerability and mortality in the age of DIY health. It is also about the ordering of letters, both of the CAG repeats, and the letters that health professionals wrote to members of my family.

10.3 Testing Times: CAG Repeats

The first reference to SCA6 came in the appointment letter, which both told my mother about the genetic results and offered her an appointment with a genetic consultant. When this arrived we looked up SCA6 online and read things about the condition. This was a distressing period, and the news was depressing but not horrific. Of the range of inherited Spinocerebellar ataxias (there are 11), SCA6 is not the worst. SCA6 at a certain point significantly impairs quality of life because people lose their balance, wobble, trip, fall and their speech becomes indistinct and slurred. It involves a degenerative loss of control over motor functions, particularly in legs, arms and throat. Wheelchairs are usually involved eventually, and it can cause death from choking or falling.

I went with my mother to this first appointment about the genetic condition. It was a tricky day, quite stressful. We found the right NHS partnership trust building and negotiated parking, none of which is straightforward. We waited in a designated area. After a time we were called into see a consultant, who told us that 'as we had been told already', nothing conclusive had been found. Somewhat confused we showed him the letter about the genetic condition which told us that something conclusive had been found. It was his turn to become confused and he asked us to go back into the waiting area.

Some time later we were called into another room. This time there was a different consultant, and some student health professionals. The consultant was enthusiastic and explained that Spinocerebellar ataxia type 6 was the issue. He explained that this was a newly identified condition and had been discovered during the course of the Human Genome Project, that it was a CAG nucleotide repeat, an autosomal dominant condition. He said that there was nothing that could be done about it, but that there was very interesting research in this area and if mum would like to join a research cohort there were opportunities for this. He mentioned the Ataxia Society as a good link should she be interested in this course of action. In the meantime, he said that she should come back every year so they could see how it was progressing. My mother did not find this to be a particularly illuminating interview but we were both reassured that we knew what was wrong and we left with an appointment for her to return in a year.

It became clear to those of us who had known my maternal grandmother that this condition had affected her, although we hadn't known that at the time. Myself, my siblings, my mother, her partner, and brother—those of us who had known Nana over a longer span of her life, and before she died, had all seen her walk around the house very shakily while holding on to the furniture and walls as she moved around. Her speech had been slurred, especially when tired or later in the day, she was difficult to understand on the phone. Eventually she had needed a wheelchair, her hands and feet shook all the time and she didn't appear to be able to speak. Still, taken as a whole, the prognosis wasn't too bad on our scale of things and it was relatively late onset. Though my mother was distressed and stressed about it, and we all felt a bit haunted by the question of its prospective presence, it didn't seem catastrophic, but on a bad day it is difficult to avoid feeling a little of its sense of doom.

My two sisters and I (all in our thirties and forties at this point) started to think about testing for ourselves, though all of the grandchildren were too young at the time. These kinds of tests are only available currently in the UK for over-18s. Of we three sisters, two of us decided to have the test and one declined. Of the two of us who agreed, my younger sister was much more proactive and clear about wanting to take the test, and signed up for it within a couple of months. She asked me to go with her when she took the test.

We were invited to a well-known genetics unit in London, where she had the test. It was interesting, and a bit odd to be involved in a process I had researched and read so much about. There were things that surprised me, though. For example, there was no reference to genetic counselling, and no one at the unit talked to my sister about what the test could mean for her children, or herself. She and I discussed the possible implications for obtaining insurance, for her children, and for her long-term health. I felt I stood in for a process of deliberation that I had imagined would be performed by health professionals. She was nevertheless clear that she wanted to know one way or the other.

The space in which we were waiting had images of socks on the walls. These images were part of an art science project. Photographic artist Gina Glover had engaged in a residency at that hospital. The socks are arranged in the shape of human chromosomes: that is, 23 pairs plus X and Y. I had read about this project, and had seen the images in other contexts, mainly in the language of grants and the sociology of biomedicine. I had been researching issues in the public engagement with genes and genetics for about five years at that point. The images were both familiar and a bit surreal in the context. My sister and I talked about them while we waited.

My sister tested positive for SCA6. I thought I was clear that I did not want to have the test, feeling the whole apparatus was an imposition. The only outcome of the test was the knowledge of a late-onset condition for which nothing could be done. I was spending a lot of time in the USA at that period of my life and was concerned about the implications for health insurance. There did not seem to be any apparent upside to having the test. However, over the course of the following year I found myself thinking about the option to have the test done. I kept deciding not to have it, but a negative decision can always be undone. The question can keep coming back. Having the test became an attractive proposition, because it would get it over and done with. Whatever the result, it would put the whole question of having to decide behind me. At least I could stop thinking about it.

A year later I also tested positive for SCA6. I was surprised at the conduct of the tests. We both saw the same genetic consultant who had advocated the tests; there was no counselling included in that process. Like my mother's consultation, there was lots of detail about the genetics but nothing much about symptoms or likely experiences, but there was absolute agreement that there is no treatment. This might explain why none of us came away from these encounters with anything like a treatment or care plan. In the case of my sister and I, there wasn't even a follow-up appointment after a year. The tests and results were just informational, it seems.

10.4 Another Ordering of Letters

When it came to obtaining my test results there were another three letters involved. These were letters from the consultant. The first letter I received offered me an appointment to discuss the condition of ataxia. The second letter informed me about the results of the test and included the DNA laboratory report. The third letter was an apology for sending me an appointment letter before I had received the results letter. The first letter rendered the results letter redundant, and resulted in the third letter apologising for the order of the first two letters. I was struck by the resonance of letters in this case where the ordering of letters, both those of the genome, and those of appointment and results are part of the choreography of diagnosis.

The experience of these letters had resonance with the experience with my mother in the first consultant meeting. The first meeting our letter gave us entry to was the one in which we were (mis)informed that nothing conclusive had been found. We could have taken that at face value, we could have accepted that mum had tested positive for SCA6 but that this didn't offer anything conclusive. In which case we might have left that interview and my mother might have gone back to her GP for further advice. This is, of course, a speculative fiction, but was a possible direction opened up by that encounter. I challenged the first consultant because the words he uttered seemed to contradict what was said in the letter, but I am not sure that my mother, on her own, would have done so.

The mistaken ordering of the letters from the consultant in relation to my test was not particularly distressing. I was not very upset about testing positive for SCA6. My motivation for taking the test was so that I could stop thinking about whether to take the test. The results confirmed a feeling of being haunted by genetics, and made me feel that genomics had become personal in a particular kind of way, even though I have as yet not had my genome sequenced. However, it was still a relief to have the question of taking the test over and done with. In another context, years before the genetic testing experience, when I was pregnant, my GP and I had had a discussion about which tests to take during pregnancy. He was usefully clear about the question of tests and he said to me at the time: 'Just because a test is available doesn't mean you should take it.' His observation highlights the normative power of testing regimes at the same time as resisting them. In the case of genomes they become normative. The question becomes why haven't you had your genome sequenced. The offer that you can have something quickly becomes a normative compulsion that you should have something.

The second of my letters, the one which properly should have been the first, wrote to me of other letters: it read: 'Kate O'Riordan has been tested for a CAG repeat allele expansion within the CACNA1A gene which is known to cause SCA6. An expansion within the affected range for SCA6 has been identified. Kate is therefore at risk of developing symptoms associated with SAC6 in the future.' It also read, 'Your gene results are now available and show that you have inherited a SCA6 gene mistake from your mother.'

The language of gene mistake and inheritance is rather more poetic than that of the CAG repeat allele expansion. It is also resonant of the language of gifts, and distributed agency. Rather than the gift of the genome, I have been gifted a gene mistake. This formulation seemed to me to create a lively sense of genetic agency, and left me wondering about the meanings of mistake and mistaken. It also made me think about mistakes along the way. If we had stuck with the first interview we had been offered when were told that nothing conclusive had been found, taken the interview as truth, rather than the letter, another unfolding of the set of events might have been different. Mistakes, taken differently, might open up a number of alternative worlds.

10.5 Life and Death Stories

My mother died on 21 December 2013 at the age of 64 from the impact of a tumour in her lung, which was already detectable via X-ray by 2009, when she was diagnosed with SCA6. After the SCA6 diagnosis, in 2009 and 2010 my mother's balance and general health did not improve, in fact it seemed to be deteriorating quite quickly, and she never really seemed to be able to shake off a cough. But we got on with work, holidays and other things, walked a 100-mile footpath called the South Downs Way over the course of several months, and went to a music festival in the summer.

It was at the music festival that mum seemed really to struggle. She was coughing a lot and became very fatigued, and on any kind of uneven ground her balance was really unstable. That made walking around a festival site pretty challenging. When we got back she went to see her doctor again, who this time immediately sent her for an X-ray. The X-ray showed a large tumour in her lung. The diagnosis this time was stage 4 terminal lung cancer with six months to live, possibly quite a bit longer if she responded positively to radiation, lung draining and chemotherapy treatments.

During the process of this diagnosis the oncology consultant accessed the X-rays that had been taken in the first round of tests, over a year earlier at this point, and these showed a significant lump in the chest cavity. Everyone was at least a bit shocked that nobody had paid attention to this information at the time. It appeared that once the genetic consultant had seen the interesting genetic result and the mistake, the results of the other tests had not been considered. This generated mistakes of a different order: mistakes in judgement, diagnosis and treatment.

Once the cancer was diagnosed we moved into a different set of experiences at a very different pace. Genomics became irrelevant, and the ataxia, although impacting on quality of life, was sufficiently overshadowed by cancer to become a curiosity, an annoyance, and a footnote. Mum was subject to a painful lung draining procedure which had to be repeated weekly, and soon after this, also to radiotherapy and chemotherapy. Her balance remained radically impaired and the treatments seemed to exacerbate the ataxia symptoms. In the meantime, she and her long-term partner organised a civil partnership ceremony in 2011. So we had a family wedding before breaking up the shared integenerational household and they moved back to the New Forest, where she had lived for 25 years previously, and had planned on returning for retirement. She stopped working when she became too ill to continue and spent the remaining part of her life in the New Forest.

10.6 Inheriting Mistakes

My mother died at the age of 64 from lung cancer, so her geneography, or genomic life story, does not play out much further than having interrupted other possible ones such as being a cancer survivor. It also prevented a story of increasing deterioration resulting from the ataxia, and the opportunity for some of us to witness our own potential futures.

Spinocerebellar Ataxia type 6 was discovered during the Human Genome Project. It remains a condition about which nothing can be done, according to conventional medical discourses. Testing for this produced different kinds of knowledge for different people. The NHS has ataxia statistics and patient records, though these are not joined-up. In a visit to my own doctor, it became evident that because of the seperation of genetic testing in the UK, there is no record of my own testing experience attached to my health records. My doctor asked me to bring in a copy of one of the letters I received over five years before so that they could update their records and see if there is any recommended follow up.

In my family we *know* that my grandmother's slurred speech, wobbling, and eventual lack of motor skills was this condition rather than the combination of tinnitus, age, and alcohol it was sometimes thought to be; the benefits of hindsight. Now I know that I will in due course start wobbling, slurring my speech, choking, and losing motor skills; the benefits of foresight. I am told nothing can be done, but of course something can be done, it depends on your take. Retaining muscle strength and overall well-being will help to offset the severity of some effects, and planning for this now seems quite a good idea. Of course, apocryphally, I might get run over by a bus. For several years, one of my sisters did not want the test, and her son, who was then over 18, thought at the time that if he had the test it would infringe her right not to know. If he tested positive she would know that she does too. At this point the grandchildren are all over or approaching an age at which they will also have a choice about the test.

It is a struggle to avoid speaking for too many others in the discourses of genomics and big data, given their inclination to aggregate and universalise. But I feel I should not try to speak for too many others, at the same time as I feel I should avoid individualization. Now I have a genomic life story, there is a 'me and the human genome'. Companies such as 23andMe, who sell genome testing, invite everyone to have 'a me and the genome' story, and many people have taken this up. I have written about this take-up for other people; conducted media analysis of genetic autobiography (O'Riordan 2011) and public understanding of genomics (O'Riordan 2016). My genome story is different from the genomic life stories I have read elsewhere. Mine does not involve interviews in *The New York Times* or science blogs, or 'spit parties' with the digerati. I will not get paid for it, nor can I claim to be part of the global science project of democratising genomes. I'm not in a research cohort or patient group, and I will not be saved. Though, it would be disingenuous to not acknowledge that I might accrue academic capital from writing about this, grist for an academic mill in which publication is supposed to prevent perishing.

In the era of DIY health care and big data we are all data subjects for whom opting out of participation is not an option. This is a media life where people are identified as data generators, and the data generated is represented back to us, so that it shapes how we understand ourselves. It is important to recognise that the aggregate data is most valuable to those orchestrating the collection at the scale of hundreds of thousands. It thus also shapes infrastructures around us, demanding protocols, processing, and code. The becoming of data subjects both exacerbates the conditions of big data and offers optimism and agency (or at least participation). This plays out unevenly, though, and information does not flow smoothly. Making health records electronic does not appear to make them any more seamless or informative. Letters still operate as the records and gatekeepers to diagnosis and treatment. Or hinder the same. Genomics does not always operate with much promise at the level of individual patients in the NHS.

Genomics England is the flagship national genome sequencing project in the UK. Despite being part of the NHS, at least in name, its high-tech visions, promises, and costs are quite separate from the everyday practice of healthcare professionals and patients in general practice. The vision of personalised medicine, imported with genomics, is one in which electronic records, data repositories, and sequence data attach to diagnosis and treatments. In correspondence with Genomics England about ataxia, they responded to my query about participation, that the genomics of ataxia are known *and therefore not relevant to them.* This mode of taking diagnosis as the end point keeps genomics at the forefront of diagnosis, tests, and data processing, but backgrounded when it comes to treatment.

Human genomics promises that we shall all have our own personal genome. This genome casts shadows, however. These can be felt in terms of lived experience. It produces effects: wonderous, exciting, and joyful for some; but evocative of fear, anger, and frustration for others. It is costly for everyone at present. The pursuit of the US$1000 genome in the USA inspired the investment of £250 million of the NHS budget in the UK in Genomics England in 2015. It also, at that time, secured a £78 million contract for Illumina, a US biotechnology company that has a monopoly on genome sequencing. At the same time, the enthusiasm for genomics, together with an ambivalence about the grounds for the project, introduce a slippage between vision and experience. In my family's experience the investment in the vision of empowering patients with information about their genomes is at odds with providing care for impairment in quality of life. The excitement of the new diverts attention and investment away from other means of experiencing healthcare. It would be worth considering to what extent that experience is writ large.

References

Johnston, I. (2016). DNA-editing breakthrough could fix "broken genes" in the brain, delay ageing and cure incurable diseases. *The Telegraph*, November 11, Wednesday.

Nuttall, N. (1991). Hunt for the Holy Grail of genetics. *The Times*, August 19, Monday.

O'Riordan, K. (2011). Writing biodigital life: Personal genomes and digital media. *Biography, 34*(1), 119–131.

O'Riordan, K. (2016). *The Genome Incorporated* (2nd ed.). London: Routledge.

Reardon, J. (2012). The democratic, anti-racist genome? Technoscience at the limits of liberalism. *Science as Culture, 21*(1), 25–48.

Zhuchenko, O., Bailey, J., Bonnen, P., Ashizawa, T., Stockton, D. W., Amos, C., et al. (1997). Autosomal dominant cerebellar ataxia (SCA6) associated with small polyglutamine expansions in the alpha 1A-voltage-dependent calcium channel. *Nature Genetics, 15*(1), 62–69.

11

Illness in the World of the Genome

Barbara Katz Rothman

This is such a sad story Kate O'Riordan is telling us—but it is one that is in so many ways just as typical of the old medicine as of its brave-new-world genomic frontiers. It is the endless tangle of blame, of medicalisation as a way out of blame, and yet somehow not a way out at all.

What do we ever ask of medicine beyond 'make it better, make it go away, fix it'? And if you cannot do that, or even if you can—remove the blame. It is one of the classic contributions of medical sociology, that medicine offers a way out of stigma. Diagnosis as a process removes responsibility—you have a disease, so it is understandable and acceptable that you missed the final exam for your course, or have not been good to your family of late, or cannot get anything done. Or walk around looking drunk. From Talcott Parsons (1951), to Eliot Friedson (1970), and right on to the work that Annemarie Jutel (2011) is doing on diagnosis, the

B. K. Rothman (✉)
Department of Sociology, City University of New York, New York, NY, USA
e-mail: BKatzRothman@gc.cuny.edu

D. Mahr and M. von Arx (eds.), *De-Sequencing*, Health, Technology and Society,
https://doi.org/10.1007/978-981-15-7728-4_11

idea of the 'patient role' is that it removes responsibility. Illness 'happens' to you; you are not to blame.

Piaget & Inhelde (1969) told us that too—we learn morality, and in so doing, we learn to care about intentions. A very young child is told the story of a little kid trying to get a cookie they were told not to have and, climbing up to reach the cookie jar, accidentally knocks into a shelf of dishes that topple over and break a cup. In another story, a little kid was trying to help set the table, and in climbing up to get the bowls, bumps into the shelf with the dishes on and breaks two cups. The little child might think breaking two cups is twice as bad as breaking one, and thus the helping kid is twice as naughty as the cookie-taking one. But an older child who is told these tales of children breaking things takes intentionality into account, as would an adult: the kid who broke one cup was being naughty and should be held accountable and punished. And the kid who was trying to help? Well, accidents happen.

But can we check into the real world for a minute? Picture yourself as the parent, or an aunt or uncle, responsible for the little kid who climbs up to get a cookie you said he could not have, and he knocks over that ugly mug your mother-in-law gave you that has been hanging around clashing with the kitchen décor for the last four years. Yes, naughty kid, yell or send him off to his room, or tell him that he will get no more cookies, or whatever it is you do to teach right and wrong. But then picture this—the little kid who decides to help set the table climbs up and knocks over a cup from the tea set your great-grandmother carried through pogroms and managed to get to New York, and your mother who died last year left it to you. Well, accidents happen? Or do you start yelling about why didn't you think, and I told you not to climb and…?

Consequences matter. It is not all about intention. And so it is with illness. Perhaps when you think that grandma's slurred speech, wobbling, and general loss of motor skills is because she drinks too much, you blame her. Or perhaps you think, well, with her bad ageing and her tinnitus, and whatever else is going on, you're not going to blame her for some drinking. Throw in a disease label to explain it, then of course, there is no blame, because there was no intentionality and it is thus not her fault. And no, no one would ever make fun of her awkwardness or be annoyed by it, or roll their eyes? Hmm. Not quite how that works in the real world. The USA just elected a President who mimicked a disabled man at a press conference.[1] Awkward gait, slurred speech, unable to do simple tasks—in the real world, those conditions, whether your fault or not, mean that life is not that much fun, and people are not necessarily all that kind about it.

And what happens when we move that diagnosis right into the definition of the person? When illness becomes not something that happens to you, but something that is part of you? 'This,' Watson famously—or apocryphally—said, holding up a CD with the genetic code written on it, 'is you.'[2] There is no 'you' to whom illness happens in the world of genome—the illness is part of what makes you who you are. That is what people with disabilities of all kinds have been saying for years: the blindness or the deafness or the wheelchair use are part of who the person is, integral to the sense of self, and not a broken bit awaiting a fix. This is what the genome reading is bringing us—not 'you have' this disease but your genome, who you are, predicts this process which we call this disease. Is what happens next very different from any other diagnosis?

Kate O'Riordan gives us the contrast right in the story—her mother's diagnosis of lung cancer. Perhaps an earlier diagnosis of that lung cancer would have made all the difference, and her mother would have been treated and cured, and lived to die peacefully at the age of 92. Or perhaps not. What we do know for sure is that she would have begun her career as a cancer patient that much earlier, and that however much this genetic diagnosis shook her up, the cancer diagnosis would have been at least as devastating. Rather than heading off on trips there would be surgeries and chemos and treatments of various sorts. Maybe she would have survived all that; and maybe not. I want to say I do not know a lot about lung cancer, but it is also true that 'they' do not know all that much either—some early-diagnosed lung cancers get treated and seem to be over and done with; but some come back, again and again until they—or the treatments—finally do kill the person.

---

[1] http://www.cnn.com/videos/tv/2015/11/26/donald-trump-mocks-reporter-with-disability-berman-sot-ac.cnn.

[2] This was so often quoted in the early days of the Human Genome Project development, but I was not able to find a citation.

Something like one in seven diagnosed lung cancers are considered 'cured' after treatment[3]—though since 'cure' is generally defined in terms of 'survival rates', any early enough diagnosis is going to look like a cure, even if one were destined to die at the same age with or without treatment. Apparently (and this is NOT based on serious, thoughtful research, just a generalisation from a handful of online sites and resources) early diagnosis can increase the cure rate by 20%.[4] Not smoking can decrease the lung cancer death rate by about 50%. Did her mother smoke? Is there anyone left who gets a diagnosis of lung cancer and does not have to explain or justify their history or non-history of smoking? Blame is powerful in many diseases, but dramatically so in this one. And yet, plenty of people die of lung cancer who had never smoked, just as plenty of smokers never get lung cancer. There is presumably a genetic component in there too.

If there is, and we find and test for it, we have a simple fix: tell people who have the genetic propensity that they should not smoke! Oh, wait, we have been telling everyone not to smoke for a very long time because it puts them at risk of lung cancer. It hasn't been all that successful a campaign. Smoking rates did eventually drop, but probably more as a result of not allowing the cigarette companies to advertise than of 16-year-olds weighing their chances of lung cancer when a friend offers a pack.

And what of the use of the genetic information in this particular case, risks not of lung cancer but of the 'gene for' SCA6? Kate O'Riordan wisely concludes that while the clinicians are telling her there is 'nothing they can do', there is no medical management, there are clearly life-style issues she can address. Stay fit, stay healthy, build up your muscles—that will make for better ageing with SCA6: 'Retaining muscle strength and overall wellbeing will help offset the severity of effects and planning for this now seems quite a good idea.'

Oh, but wait: that is what makes for better ageing for everyone. Don't we tell everyone to exercise, maintain muscle strength, take overall good

---

[3]https://health.clevelandclinic.org/2012/11/defeating-lung-cancer-with-early-detection.

[4]https://www.iaslc.org/lung-cancer-screening-and-early-detection-can-increase-curability-and-save-lives.

care of themselves to age well and be healthy. So, like those with some genetic propensity towards lung cancer, we give the same advice we give everyone and add 'And this time I mean it!' Like we did not really mean it all the other times we tell them not to smoke, to eat right, and to exercise.

So here we are in the brave new world: predictions that will, of course, be relatively vague, just how severe or just how early-onset is not known. When the symptoms come, you will have a medical diagnosis, which should, we think, relieve you of responsibility even if it cannot cure you. But it won't. Predictions, even less than diseases, do not relieve of us responsibility, but if anything focus and increase that responsibility: Did you exercise enough? And even if you did, even if you are not being held 'responsible', well… Life's not easy with a disability. The stigma remains.

References

Friedson, E. (1970). *Profession of Medicine: A Study in Applied Knowledge*. Chicago, IL: University of Chicago Press.

Jutel, A. (2011). *Putting a Name to It: Diagnosis in Contemporary Society*. Baltimore, MD: Johns Hopkins University Press.

Parson, T. (1951). *The Social System*. New York, NY: Free Press.

Piaget, J., & Inhelde, B. (1969). *The Psychology of the Child*. New York, NY: Basic Books.

12

How Personal Is the Genome? The Shadow of Genetic Predictions

Silja Samerski

With big data and predictive analytics, fortune-telling has become a new service product. From digital epidemiology to genetic testing, from election forecasting to calculating repeat offence risks—anticipated futures overshadow our present. Living in the *hic and nunc* seems outdated; instead, 'anticipatory action' has become an imperative, in business matters as well as in daily living. Get hold of the future! Aside from business and 'security', 'health' is the main promise that drives data-based predictions: doctors offer various tests, from mammography to genetic testing, to calculate health risks, and people track their heartbeats, temperature, and fitness status to detect risk factors and optimise their health performance. Thus stories, such as Kate O'Riordan's, of being burdened with a disease prediction is peculiar to our time.

S. Samerski (✉)
Social Work and Health, University of Applied Sciences Emden/Leer, Emden, Germany
e-mail: silja.samerski@hs-emden-leer.de

D. Mahr and M. von Arx (eds.), *De-Sequencing*, Health, Technology and Society,
https://doi.org/10.1007/978-981-15-7728-4_12

12.1 Futures

Foresight is always speculative. No one foresaw that Kate O'Riordan's mother would die from lung cancer only a short time after she was diagnosed with a genetic disorder. No one can know if, when, and how Kate will experience the first symptoms of ataxia. No matter if a predicted scenario is calculated by a statistician or by the algorithms of IBM's supercomputer Watson, and no matter if it is deduced from big data or from genetic data—individual predictions inevitably remain probabilistic. Personal data can only be interpreted in the light of statistics: even the most personalised risk profile needs reference to population statistics to be more than merely a database of random information. This holds true for vital data as well as for genetic tests. Neither a molecular marker nor a sequence of DNA of itself reveals anything about a person's state or future. As geneticist Richard Lewontin clarifies, genotype and phenotype can be correlated, but it is not possible to extrapolate from genotype to phenotype in a single case: 'As is true for living systems in general, relations between genotype and phenotype are contingent, varying from case to case' (Lewontin 2004). Rather, a genetic test enables statisticians to reduce a testee to one more case they can add to a stigmatised population, just as a genetic fingerprint adds an accused to the population of criminal suspects. But while those who are innocently accused know they need not feel any guilt, anyone who tests positive for a genetic condition will from then onwards be dis-eased, regardless of the validity of the test.

Kate O'Riordan's story about her mother's last year and about her family struggling with genetic testing options and test results is, above all, touching. Through her sober narration, the reader senses the author's worries and perplexities as well as the closeness between mother and daughter. Her story is a personal narrative but at the same time sheds light on the prediction industry obsessed with collecting data and calculating speculative futures. As Kate describes, the genetic tests had several effects on her family and on their relations with one another. After her mother was tested, Kate suddenly saw and understood her family's biographies in the light of 'the genome'—an abstraction we never experience but are only told of by experts. Through the reference to the genome, her grandmother's awkwardness and her mother's vertigo were given a professional label and explanation. More than that: what before might have been interpreted as a personal oddity or affliction was reduced by fiat to a biological error; to a kind of biochemical spelling mistake that has always been there, everywhere in the whole person. Thus reference to the genome fundamentally changes the status of a disease: unlike the cases of dementia, liver cirrhosis, and muscle weakness, genes and gene mutations are not something someone has; rather, they are what someone *is*. The attribution of 'something genetic' is thus much more powerful than a conventional medical diagnosis. Genetic tests imply that geneticists shed light on the invisible truth about one's self. They reveal a hitherto undetected 'identity': 'Genetic disease differs in as much as it promises to reveal who the individual always has been, not a new addition but a revelation about an underlying identity that had been concealed' (Armstrong et al. 1998: 1658; see also Samerski 2015). Thus, as soon as predictions are based on genetic data, fortune-telling gells into purported truth telling.

12.2 How Personal Is the Genome?

While the mutation SCA6 in her mother's and grandmother's case offers a biomedical interpretation, of what is and what has been, of their suffering, it has a different effect on Kate herself: in her case, the test result is not an explanation of the present or past, but rather a verdict. Her future, the test suggests, is not open and revealed by time, but is already there: programmed in her DNA. The biggest health threats, genomics suggests, do not come from outside, but from inside, and are part of one's system. Future calamities lurk as faulty genes in our interior. This image of a hidden threat from within dovetails with today's fear of terrorism and with security practices of the twenty-first century: in a constant battle with mutations and errors, the body is understood as a kind of surveillance system. A biologist makes this analogy with anti-terrorist security explicit: genes are like 'terrorists', biologist John Turner writes: 'They have the power to kill, maim, or make life downright miserable for us and our children. Some strike at birth, others "sleep" for decades, and, like good terrorists, they are so well integrated into our body politic that, until the last few years, their exact whereabouts were a mystery: their individual extirpation (or more properly correction) is still well nigh impossible' (Turner 2001: 8).

Yet the genome is neither an enemy from within, nor the book of life that reveals people's destinies. Rather, it is an aggregation of data from which statisticians generate risk profiles and model probable futures. In order to make a biochemical sequence predictive, it needs reference to populations. However, this statistical rationality stipulates a form of cohort thinking through which the individual citizen is turned into a faceless member of a totality. Predictive and preventive strategies, sociologist Thomas Lemke writes, dissolve the notion of the subject that is substituted by an aggregate of factors, risk factors (Kollek and Lemke 2008: 159). Thus the genome entangles us in peculiar contradictions: It promises knowledge and predicts uncertain futures; it facilitates verdicts and calls for self-optimization; it connotes intimacy and transforms testees into faceless members of populations; it is a statistical construct and claims to be personal (Samerski 2015).

These paradoxes, however, become less and less apparent in a society in which big data, digital surveillance, and predictive analytics invite people to see and treat themselves as data profiles. As Emily Martin has shown, social and economic demands coincide with the scientifically transmitted view of the body in an unsettling way. In the 1990s, the immune system was a dominant metaphor that adapted people's self-perception to the basic concepts and images of postindustrial society. With the immune system, she explains, 'people are, so to speak, coming to see themselves as mini-corporations, collections as assets that each person must continually invest in, nurture, manage, and develop' (Martin 1994: 77). Today it is rather the genome that redefines the body as a risk-laden system with internal threats. This reinterpretation of the body as a flexible system that needs surveillance and management makes it seem plausible to accept the laws of statistical cohorts and probability distributions as one's own. The attributed risks, counted for populations, seem to measure the threats from within. Today's body image, heavily influenced by genetics and information technology (IT), makes us more or less physically compatible with predictive analytics. Yet, as Kate's narrative shows, the future remains unknown. Not to be overwhelmed by the shadow of forecasts will be one of the main challenges to well-being in a time of data-driven predictions.

References

Armstrong, D., Michie, S., & Marteau, T. M. (1998). Revealed identity: A study of the process of genetic counselling. *Social Science and Medicine, 47*(11), 1653–1658.

Kollek, R., & Lemke T. (2008). *Der medizinische Blick in die Zukunft. Gesellschaftliche Implikationen prädiktiver Gentests*. Frankfurt a.M.: Campus.

Martin, E. (1994). *Flexible Bodies. Tracking Immunity in American Culture—From the Days of Polio to the Age of AIDS*. Boston, MA: Beacon Press.

Lewontin, R (2004). *The Genotype/Phenotype Distinction*. The Stanford Encyclopedia of Philosophy. http://plato.stanford.edu/entries/genotype-phenotype/, 12 February 2010.

Samerski, S. (2015). *The Decision Trap: Genetic Education and Its Social Consequences*. Exeter, Devon: Imprint Academic.

Turner, J. R. G. (2001). Self-made men (review). In Times Literary Supplement, 5142.

Part V

Individual Experiences

13

Lived Genome Phenomenology: Exploring the Genetics of Chronic Inflammatory Bowel Diseases

Dana Mahr

13.1 Introduction

Chronic inflammatory bowel diseases (IBD) have recently been reshaped as clinical conditions with a genetically determined susceptibility, adding genetic factors to the explanation of chronic autoimmune conditions of ultimately unknown origin (Schreiber et al. 2005). This shift emerged from the predominance of molecular and genetic approaches in the life sciences and resulted in a drastic reconfiguration of IBDs: from a psychosomatic device, to a multifactorial collage of symptoms within the episteme of genetic susceptibility (Mahr 2017). IBD patients born in the 1960s, 1970s, or 1980s, have been socialised within the psychosomatic paradigm, yet in the contemporary medical system they have to cope with genetic explanations. Against this backdrop, I ask two questions. The first is of a phenomenological nature: How do patients and their families make sense of the shift within the aetiology of their disease

D. Mahr (✉)
Section de Biologie, University of Geneva, Geneva, Switzerland
e-mail: dana.mahr@unige.ch

D. Mahr and M. von Arx (eds.), *De-Sequencing*, Health, Technology and Society,
https://doi.org/10.1007/978-981-15-7728-4_13

in their individual life worlds? The second question has a more analytical focus: How can their experience at the intersections of embodiment, social configurations, and molecular genetics enrich the interdisciplinary programme of 'de-sequencing'?

In the first section of this chapter I shall develop a sociologically informed framework to understand the highly individual processes undergone by people suffering from IBD or other genetic diseases. I understand it as a 'lived genome phenomenology' within a 'geneticised' world. In the second (genetics as a relief) and third (genetics and agency) parts of the chapter, I shall explore and analyse two complementary 'first-hand' sense-making narratives of IBD patients, and in the final section, I seek to disentangle such narratives and re-integrate them into the broader discussion of personal genetics in the twenty-first century.

13.2 A Phenomenological Approach to Genomic Knowledge

In molecular biology and medicine, the term 'genome' describes the complete genetic material of an organism, all the genetic information that actually 'makes' an individual. In societies such as Europe and the USA, in which genetic tests are increasingly becoming a standard tool of clinical research, practice and even lifestyle (in the form of direct-to-consumer genetic testing), the genome has furthermore become an increasingly important social actant (Latour 2003). Questions arise, such as: what does my genome tell me and others (e.g. family members, employers, health insurance companies) about myself? Am I my genes (see Klitzman 2012)? Do my genes determine my destiny, and that of my family?

In recent years, bioethicists, together with sociologists and other social scientists, have spilled much ink in contextualising, sorting, and answering these questions. Within this discourse, the term 'geneticisation' has attracted much attention as a concept describing the possible and actual impact of genetic knowledge on modern societies and the individuals living in them. The geneticist and feminist scholar, Abby Lippman, introduced the term 'geneticisation' in 1991. For her, it 'refers to an ongoing process by which differences between individuals are reduced to the DNA codes, with most disorders, behaviours and physiological variations defined, at least in part, as genetic in origin' (Lippman 1991: 19). Differences between individuals that are redefined as genetic can nevertheless be studied sociologically. Some of these differences might be lost in this reductionist view, as might the degree to which these variations acquire meaning in the social spheres of those actually living them (Rehmann-Sutter and Mahr 2016: 93). But in the past the discourse of geneticisation centred mainly on aspects of risk and the management of risk, either at the reproductive level of family and kinship, as analysed by Kathie Featherstone et al. (2006), Kaja Finkler (2000), Monica Konrad (2005) and Andrea zur Nieden (2014), or at the level of society and health governance, as in the work of Andrea. The genome has furthermore become an increasingly important social actant, as Lemke shows (2004, 2006a, b). The social dimensions of genetic knowledge and its meaning are portrayed ambiguously in these accounts.

To explore whether and how genetic knowledge can also be framed as positively meaningful, it is vital to expand our understanding and conceptualisation of geneticisation. I argue that, instead of framing genetics as reductionist and inherently oppressive, I first have to identify the social questions raised by the implementation of genetic knowledge and explanations of disease into everyday lives (Rehmann-Sutter and Mahr 2016). Furthermore, it is necessary to examine more closely the micro-sociological level of illness rather than looking into distant dimensions of time and space, or at the speculative level of shifts in whole societies (the bio-sociality of genetic knowledge). Therefore, I suggest the study of how genomic knowledge is individually lived. What hopes and fears do people direct towards genetic explanations for diseases, even before they have specific knowledge about them? What are the benefits of genetic explanations for health conditions, and what are their disadvantages? How are genetic knowledge and genetic explanations of diseases integrated into (or complemented by) explanations that are not primarily genetic? Questions such as these underpin phenomenological and qualitative research perspectives, since they centre on both discursive and tangible practices of individual understanding. Hence, individual narratives of those individuals who are experiencing the shift towards a genetic explanation of their own condition are important sources for a better understanding of how genetic knowledge colonises and transforms our everyday life.

To study such narratives from a phenomenological perspective, I propose five interpretatory essentials. First, the exploration of genomic sense-making needs to abstain from value judgements towards how patients interpret their genes or the genome. (1) The epistemic object called 'the genome' that scientists and doctors analyse and discuss should have no higher authority than the most humble subjective conceptualisation of a patient. Because her, his, or their, first-person description is a valid source for understanding this kind of entity from a microsociological perspective. (2) The sociological place of knowledge is always a practical one. To observe and understand it, it is necessary to analyse human behaviour and activities in the context of everyday life. (3) A crucial point is that individuals and small groups such as families need to be the centre of exploration, since a lived perspective on genomic knowledge can best be understood through the unique ways in which individuals experience, reflect, and explain diseases in their individual lives. (4) The phenomenological perspective seeks to make sense of the sense-making of others. Lived experience is what a phenomenologist gathers, rather than traditional quantitative data. (5) Phenomenology is about discovery. For this, qualitative (and non-restrictive) methods should be the tool of choice.

Crohn's disease and ulcerative colitis are widespread throughout Western society and have been re-described several times over recent decades. In many European countries, the USA, and Canada, large self-help organisations have existed for IBD since the 1980s. Among others: 'Crohn's and Colitis UK', the 'Crohn's and Colitis Foundation of America', and the 'Deutsche Morbus Crohn/Colitis ulcerosa Vereinigung' (DCCV) (Koay and Sharp 2013). With help from the DCCV, support groups in social networks and 'tribes' in social health networks such as PatientsLikeMe, I invited patients and families to participate in a series of semi-structured interviews. In the initial call for participation I asked them to talk about how they led their lives with an IBD, and how they understand the fact that their condition is now interpreted within the paradigm of genetic risk susceptibility. For this, I planned 1–2 interviews per participant with a length of about 1 hour. I invited potential participants to choose their own location for the interviews, to give them the opportunity to talk about their highly personal and sometimes difficult experiences in a safe space. A total of 81 people agreed to participate in my study. Out of this group, I interviewed 41 persons once, two times at an interval of several months. I wanted to represent in this sample participants from all layers of society, both genders (with a female to male bias of 3–1), and ages ranging between 16 and 65. I included the voices of both patients and their family members (spouses, life partners, parents, children, cousins, etc.) to explore sense-making processes from different angles (Mahr et al. 2019). All interviews used for this publication were professionally transcribed and anonymised. The names used are pseudonyms.

13.3 Genetics as a Relief

Since the current discourse on geneticisation centres mainly on the individual and/or societal dangers of developments in genetic research and diagnostics—such as the social toll of living with a genetic 'predisposition' (Konrad 2005), or a discussion of the scientific generation of 'presymptomatic individuals' whose social identities are both healthy and simultaneously endangered—I was surprised at our phenomenological findings in the narratives of IBD sufferers.

The specific ways in which scientists, doctors and healthcare professionals frame IBD influence both treatment regimes and how patients perceive and manage their social lives and lived bodies. The dominant aetiology of IBD has altered at least twice in recent decades—from a psychosomatic explanation in the 1970s and 1980s, to the genetic risk paradigm since the 1990s, plus the complementary approach of microbiomics—and this means that patients and their families have had to undergo the same changes in their own lives. In this process, IBD patients also become subjects and interpreters of the medical history of their own disease. However, their disease histories are far more experience-laden and come from a different angle than those of professionals. This proves to be a rich source of knowledge.

In the 1980s, when leading scholars such as Hubert Feiereis framed Crohn's disease and ulcerative colitis as psychosomatic conditions, they also 'made' or 'formed' the social identities of IBD sufferers—for example, by defining their typical personality structures and by searching for psychological causes that triggered the development or outset of the condition (Leibig et al. 2003). This always came with a 'social toll', since IBD patients were often framed as being 'mentally weak' or not 'resilient' within (and by) their social environments (Gazzard et al. 1978). Elsewhere, we have shown how this led to a feeling of personal guilt in the lives of many patients (Rehmann-Sutter and Mahr 2016). As the 65-year-old study participant, Agathe, told us, for these patients the psychosomatic explanation of IBD could be understood as 'a "my-fault theory". It blames the disease on the patient, saying something like: "You (just) have to try hard enough and you'll be well again"' (Interview with Agathe). This meant that the patients themselves were deemed to be the disease-causing factor, in terms of psycho-hygienic self-care.

The phenomenological exploration of patients' narratives led to the insight that, against this background (and beyond its scientific value, medical options, or risk potentials), a central meaning of the genetic explanation within the lives of many IBD patients is its potential to counter the implicit social blame connected to the psychosomatic explanatory model. This is particularly the case for those patients who were over 25 years old and had been socialised within the psychosomatic explanatory model. For them, the genetic explanation is a relief, since it 'transcends' the cause of the disease from the subjective space of the patient. It is no longer 'me' who has failed but the 'immutable' and 'fate determining' entity called the genome. Although this reductionist picture is no longer perpetuated within systems biology as a general explanatory frame for genetics, it is still a meaningful aspect of a patient's sense-making strategy. In the words of Nina, a 47-year-old medical secretary and Crohn's patient whom I interviewed in April 2014:

> Yes, I think it's really great that it's genetic, or at least 50 percent of it is … Because people say 'You had a relapse? Oh, were you stressed again?' … So

> I was questioning myself all the time ...This psychosomatic explanation always made me feel that I'm crazy, I can't cope with my everyday life – my God, what a loser I am ... So in my eyes, it [the genetic explanation] is a real liberation. (Interview with Nina and her husband Jens)

Kerstin, a 43-year-old interviewee, whose severe form of Crohn's disease makes her unable to work, re-interprets her whole illness story (and that of her family) in the light of the genetics of IBD:

> They always say, yes, uhm... the bowel... it's psychosomatic. But I had a very happy childhood. I could not say that something went wrong in my childhood, that something influenced me or caused me any problems. I cannot remember anything like that... Nothing dramatic, a normal life... really nothing... that one could say OK... it's the mind that is the problem. But I have to say that my father and my sister often had abdominal pain and diarrhoea. But my father... had a colostomy... there's nothing. But there is really also the side [of the family] [to which] my second cousin and aunt belong. So I think maybe it has to do with my genes. That the genes bring the bowel problem. (Interview with Kerstin)

Kerstin and Nina, whom I have quoted as representative of a group of 18 other individuals who provided similar perspectives, frame the genetic explanation of IBD as a tool for self-empowerment against the social struggles connected with the older, psychosomatic interpretation. For them the genetic explanation has a real, lived meaning. It can be used in specific social situations and has the power to re-frame their whole illness experience and history—as we have seen in the quote from Kerstin.

This brief insight into phenomenological work on individual meaning-making in chronic illness shows that the study of geneticisation needs to be broadened. Beyond the problem-centred approaches of contemporary discourse in bioethics and sociology, my example suggests that the social processes of geneticisation may hold positive meanings for specific populations or in specific situations. But it should be emphasised that (beyond its medical or scientific importance) the awareness of 'living a genome' is not always present in people's heads. Almost always it needs to be embedded in a process of reflexive embodiment (Mahr et al., forthcoming). Since we cannot feel or sense our genome directly (as we can, for example, our pulse) it seems that we often need concrete experiences on another level (such as the struggle with an alternative medical explanation) to prompt us to reflect on its meaning. Furthermore, this process of reflexivity needs to promise some kind of change, somatically (e.g., therapy), socially, or in our self-understanding. In other words, the genetic knowledge needs to have an impact on lived lives.

13.4 Genetics and Agency

Complementary to the large group of participants who interpreted the genetic explanation as a relief since it helps them to 'counter' negative social attributions of IBD sufferers (here represented by Kerstin and Nina), there exists a second group which is more sceptical about the genetic classification of their condition. For them, the genetic explanation absolves patients of their personal responsibility towards their disease. This group of interviewees prefer the older, psychosomatic model, since they tend to find in it more ways to act autonomously. In contrast to people such as Kerstin and Nina, they received positive feedback from their social worlds or have proved to be very resilient against guilt narratives. Therefore they do not find the 'relief function' of the genetic explanation relevant or helpful.

Sabine, aged 45, gives a good example of this. For her, 'being responsible for oneself' (Interview with Sabine) is a *conditio sine qua non* for having a fulfilled life while coping with IBD, because in her case psychotherapy alongside other practices of psychological self-care had a major impact on her well-being. The genetic paradigm of IBD seems to threaten this perspective, since it makes it far too easy to diffuse individual responsibility (Rehmann-Sutter and Mahr 2016). Sabine explained to us:

> I think general personal responsibility gets a very raw deal, and uhm, if our musicians [the scientists and doctors who advocate genetic explanations for IBD – DM] now start to sing its praises [of the genetic risk paradigm – DM], uhm, then I'm, that's my fault alone. So I would agree, that it's perhaps ignorance, it's definitely terribly frightening to say all at

> once that there's something physically wrong and I'm responsible for it myself. (Interview with Sabine)

In contrast to people like Kerstin, Nina or Agathe, our interview partner Sabine does not interpret Crohn's disease or ulcerative colitis exclusively in terms of an inevitable 'genetic destiny' but as something that—despite its flaws—can also acquire a positive function in one's life. She even speaks of her 'Crohn's' as if it was a person and interprets it as a companion in self-care. For example, 'he' warns her when she is demanding too much from her body or does not listen to less painful and less problematic stress signals:

> I am very strongly convinced that, uhm, Crohn's is my buddy, that he's saying inside me: Now look, you've gone too far here. Or here, stop do this or that… I don't know. I have read a wonderful book from Rüdiger Dahlke, Illness as Symbol [Dahlke 2007] and accepted finally that I'm a 'slime bag' [German: Schleimscheisser]. (Interview with Sabine)

At first glance it seems commonplace for diseases and the explanatory models surrounding them to gain meaning in the experienced lives of patients. But the applied phenomenological approach we provide here offers the possibility of exploring more deeply. As this example of the geneticisation of IBD already shows, (1) it helps to gain information about the different social spheres and social experiences in which explanatory models or scientific concepts acquire individual and social meaning; (2) it allows us to investigate how (and under what circumstances) they are utilised by patients; and (3) it also enables us to see how they are translated into these processes of individual sense-making. I was able to accompany Kerstin, Nina and Agathe, whose bad experiences with the former explanatory model of IBD led them to embrace the genetic model, since they interpreted it as a tool to counter negative attributions in their daily lives. But I was also able to show how another group (with a different basic structure of social experiences—here represented by Sabine) developed a complementary and more critical stance towards the genetics of IBD. In the view of this group of IBD patients, genetic knowledge about Crohn's or colitis carries the risk of reducing self-responsibility and the motivation for self-care—because one can easily say 'It's not me but my genes.' This view displays the fear that individuals could become discouraged from being, or becoming, active in their situation because of the 'fate-like' character of genetics.

It is also interesting to notice that (despite differences in the individual use of explanations and the value attributed to the genetic model) almost everyone I interviewed had a strong and robust concept of the entity that we call a 'gene'. In their narratives, a gene was portrayed as an essentialist object. Therefore they (still) often characterised genes as something that determines 'who we really are'—following the still prevailing pop cultural imaginations about genes displayed, for example, in the movie 'Gattaca'. That is why genetic information about IBD can at the same time be both a burden and a relief. Multifactorial and less reductionist views such as epigenetics are not yet part of first-hand perspectives about genomic knowledge.

13.5 Sense-Making as a Form of De-sequencing

Genetic knowledge gains social significance through the interpretation of lived experiences. Unlike in science, where genetic knowledge has an epistemic value in itself, its significance in concrete lives is highly dependent on a social *raison d'être*. This means that this 'knowledge coded in us' can only be experienced in everyday life under two conditions: either when it is interpreted for us by experts and we are asked to base our life decisions on it, or when we reinterpret our bodies, our social relationships and our possible futures against the background of such knowledge. In a dialectical way, an epistemic object is taken out of its original context and transformed into an element of interpretive power within individual lives. The abstract representative form of DNA is dissolved and re-sequenced in the life worlds of individuals with IBD. The instrument of this transformation is not the genetic scissors of CRISPR/Cas9 technology but an experience-based search for meaning (Mahr et al. 2019). This form of social transcription tends to appear open to interpretation, but is not arbitrary. In my study of the experiences of people with IBD, I have identified a continuum with two interpretative poles around which the attributions of meaning and social positioning of my interview partners are formed. The genetic explanations of diseases such as Crohn's disease or ulcerative colitis are perceived as something positive and liberating at the level of individuals when their social experiences of living with their disease have been rather negative. The IBD genome is then transcribed into self-apologetic actions and abstracted in a new way: 'It's not me who is to blame for my condition, but my genes.' In other words, through experience, something epistemic is transformed into something moral. Consequently, genes are understood here as a rather deterministic quantity. The other pole of meaningful interpretations in my research is characterised by persons whom I would like to describe as 'gene-relativist' self-improvers. Their re-interpretation of genetic information is characterised by a mimicry of the scientific context in which such knowledge originates. They accept that a genetic defect is partly responsible for their physical condition, but they also know that the associated disease affects their lives in a multifactorial way. Especially when the social experience of living with IBD is interpreted as positive, i.e., when relatives and other caregivers are perceived as being helpful rather than antagonistic, then the genetic explanation of IBD is perceived as being less significant. This shows that the individual social contexts in which genetic information is inscribed have a high sociological relevance. For it is these contexts that give genes the power to influence the individual life plans of a person in the long term. How will the geneticisation of more and more diseases affect our family planning in the future? Not only genetic counsellors should pay attention to this, but also social and health policy-makers and the scientific community. Now, as the other contributions in this volume show, is the time when we are in a position to study the social impact of genetic knowledge in 'real time'.

References

Dahlke, R. (2007). *Krankheit als Symbol. Handbuch der Psychosomatik*. Gütersloh: C. Bertelsmann.

Featherstone, K., Atkinson, P., Bharadwaj, A., & Clarke, A. (2006). *Risky Relations: Family, Kinship and the New Genetics*. Oxford: Berg.

Finkler, K. (2000). *Experiencing the New Genetics: Family and Kinship on the Medical Frontier*. Philadelphia: University of Pennsylvania Press.

Gazzard, B. G., Price, H. L., Libby, G. W., & Dawson, A. M. (1978). The social toll of Crohn's disease. *British Medical Journal, 2*(6145), 1117–1119.

Klitzman, R. (2012). *Am I My Genes? Confronting Fate and Family Secrets in the Age of Genetic Testing*. Oxford and New York: Oxford University Press.

Lippman, A. (1991). Prenatal genetic testing and screening: Constructing needs and reinforcing inequities. *American Journal of Law and Medicine, 17*(1–2), 15–50.

Koay, P. P., & Sharp, R. R. (2013). The role of patient advocacy organizations in shaping genomic science. *Annual Review of Genomics and Human Genetics, 14*(1), 579–595.

Konrad, M. (2005). *Narrating the New Predictive Genetics: Ethics, Ethnography, and Science*. Cambridge Studies in Society and the Life Sciences. Cambridge, UK and New York: Cambridge University Press.

Latour, B. (2003). *Science in Action: How to Follow Scientists and Engineers Through Society*. 11. print. Cambridge, MA: Harvard University Press.

Leibig, T., Eberhard, W., & Feiereis, H. (2003). Zur Persönlichkeitsstruktur von Patienten mit Colitis Ulcerosa und Morbus Crohn, Eine Testpsychologische Untersuchung während der Krankheitsremission. In *Zeitschrift für Psychosomatische Medizin und Psychoanalyse* (pp 380–392).

Lemke, T. (2004). Disposition and determinism—Genetic diagnostics in risk society. *The Sociological Review, 52*(4), 550–566.

Lemke, T. (2006a). Lebenspolitik und Biomoral: Dimensionen Genetischer Verantwortung. In *Die Verfasstheit der Wissensgesellschaft* (pp. 333–345). Westfälisches Dampfboot.

Lemke, T. (2006b). Genetic responsibility and neo-liberal governmentality: Medical diagnosis as moral technology. In A. Beaulieu, & D. A. Gabbard (Eds.), *Michel Foucault and Power Today: International Multidisciplinary Studies in the History of the Present* (pp. 83–91). Lanham: Lexington Books.

Mahr, D. (2017). Narrating genes: How patients with chronic inflammatory bowel diseases interpret an emerging disease aetiology and how we can make sense out of it by developing a historically and sociologically informed framework. In *History of Human Genetics* (pp. 325–336). Cham: Springer.

Mahr, D., Mahr, E., & Rehmann-Sutter, C. (2019). Subjektivierungsfiguren genetischer Information. *Sozialer Sinn, 20*(1), 1–39.

Rehmann-Sutter, C., & Mahr, D. (2016). The lived genome. In *Edinburgh Companion to the Critical Medical Humanities* (pp. 87–103). Edinburgh: Edinburgh University Press.

Schreiber, S., Rosenstiel, P., Albrecht, M., Hampe, J., & Krawczak, M. (2005). Genetics of Crohn disease, an archetypal inflammatory barrier disease. *Nature Reviews Genetics, 6*(5), 376–388. https://doi.org/10.1038/nrg1607.

Zur Nieden, A. (2014). Zum Subjekt der Gene werden Subjektivierungsweisen im Zeichen der Genetisierung von Brustkrebs, Transcript, Bielefeld.

Primary Sources

Interviews with Agathe, Nina and Jens, Kerstin, Jonas, Sabine, Elfriede, Esmeralda, and Marianne.

14

Existential Storytelling in a Genomic World

Stephan Kopsieker

In her contribution to this volume, Dana Mahr invites us to broaden our understanding of the 'geneticisation' of Western society by offering a micro-sociological and phenomenological perspective on the impact of genomic knowledge on individual biographies. Mahr presents some very interesting findings about how people deal with chronic diseases in utilising (or opposing) a certain kind of knowledge for telling their biographical stories. This 'genomic knowledge' becomes available through the scientific (re-) interpretation of their chronic diseases as 'genetic'.

As Mahr tells us, the concept of geneticisation is intended to capture the impact of genetic knowledge on Western societies and the individuals living in them. The perspective on geneticisation was hitherto mainly concerned with risk and management of risk at the level of family or society and health governance. But Mahr shifts the perspective to the level of the individual and puts the focus on the generation of meaning in individual narratives, which are impacted by genomic knowledge. In this

S. Kopsieker (✉)
Brill Deutschland GmbH, Paderborn, Germany
e-mail: kopsieker@brill.com

D. Mahr and M. von Arx (eds.), *De-Sequencing*, Health, Technology and Society,
https://doi.org/10.1007/978-981-15-7728-4_14

context, the process of geneticisation could be understood as describing the growing influence of genomic knowledge on the way people create meaningful stories about their lives.

The phenomenological approach used by Mahr reveals many interesting findings that could be discussed at length from a philosophical viewpoint. The relation between epistemic resources and action, the interplay of knowledge and emotion, or ethical questions concerning the responsibility of researchers, who are influencing the lives of their human subjects, are just a few of the topics touched on in Mahr's contribution, and would deserve a more lengthy philosophical treatment.

In this commentary I shall focus on one aspect of Mahr's study that concerns the kind of storytelling that people with chronic diseases use to interpret their social and existential situation. For lack of a better term, I will call this kind of storytelling 'existential storytelling'. The practice of existential storytelling illustrates very nicely the intimate connection between phenomenological approaches and philosophical accounts that are gathered under the label of existentialism.

14.1 Existential Storytelling

Mahr explores individual narratives of people who have to deal with the complexities of making sense of living with a chronic disease, and the incorporation of their genomic knowledge into their narrative. The working hypothesis is that the analysis of these individual narratives can be used eventually to broaden our understanding of the role of genetics in modern Western societies.

The term 'individual narrative' refers to the stories people tell about themselves that are intended to give a meaningful interpretation of their current situation, their disease, their expectations, and their wishes for the future. These stories incorporate the many events that make up a person's life, their knowledge and beliefs, and their wants and needs. The individual narratives people tell about themselves are the product of a practice that could be called existential storytelling.

Storytelling as a device to produce meaning has a prominent role in many philosophical positions that are often regarded as existential and phenomenological, e.g., in the works of Kierkegaard, Sartre, Beauvoir, Merleau-Ponty, or even the political philosophy of Hannah Arendt. All of these philosophers have taken into account the lived experiences of humans and their existential conditions. The existential and phenomenological influence in Arendt's work becomes especially notable in *The Human Condition* (Arendt 1958). For Arendt, storytelling—that is, compiling the actions and statements of an individual into a narrative—is constitutive for the meaning of a single action or statement. Only in a retrospective articulation can the significance and import of an action or statement be assessed (see d'Entreves 2014).

Furthermore, some philosophical accounts of personal identity emphasise the role of storytelling and biography, and make use of the so-called 'narrative criterion of personal identity'. According to this criterion, a characteristic, action, or experience is attributable to some person by its incorporation into the self-told story of his or her life. This narrative identity builds the kind of psychological unity that is needed to identify an individual. The coherence and intelligibility generated by the individual narrative are also intimately related to the creation of meaning. Schechtman (1996) claims that only in light of these unifying self-told stories can actions and events acquire real meaning. Rather than merely being isolated events, the larger story relates experiences to one another within the context of one life (see Shoemaker 2016). It would certainly be interesting to explore the connections between storytelling, personal identity, and the production of meaning in more detail, but I want to point out another connection between the practice of storytelling and existential philosophy.

Kierkegaard, Sartre, and other existential philosophers have emphasised the importance of subjective stories and the practice of storytelling for dealing with existential issues (see Kenyon [1996] for a more detailed account of the relation between storytelling and existential philosophy). The practice of dealing with an existential crisis by creating a meaningful story about one's life and the way one experiences it is what I want to call 'existential storytelling'. In Mahr's study we can observe the practice of existential storytelling in action. Suffering from a chronic disease for which there is no permanent treatment certainly counts as some form of existential crisis. So people have to make sense of their situation and integrate their experiences, actions, and statements about it in a meaningful view of the world and their own respective place in it. I would like to highlight some interesting features of this practice of subjective storytelling and the generation of meaning that have become evident through Mahr's observations.

One thing that has become clear is that the practice of existential storytelling involves the use of different resources to construct a meaningful narrative and an interpretation scheme for an indivdual's own (his)story. With the generation of new scientific knowledge about the condition of IBD, a new resource for stories of those who suffer from it becomes available: genomic knowledge. By shifting the type of explanation from psychosomatic to genetic, the genomic knowledge has the potential to impact the individual biographies of those suffering from the condition. Patients might experience their illness in a new way, or make adjustments to their self-told stories. Since their lives are influenced directly by the new genetic explanation, it could be said that their life-world is now a 'genomic world'. As I would put it, a person lives in a genomic world, when genomic knowledge has a potential influence on his or her lived experiences, self-depiction, and actions.

In this sense, Mahr's phenomenological study aims to connect the significance of genomic knowledge 'From a phenomenological perspective, the significance of genetic knowledge is connected to the individual meaning it acquires in the actual lives of those who suffer from a condition that becomes geneticized'. One way that the genetic explanation of IBD becomes significant is as a new (potential) resource for existential storytelling. At the same time, a resource is lost, since IBD is no longer explained as a psychosomatic disease. Furthermore, Mahr's study shows that the availability of a certain resource can have different effects on individual stories.

14.2 Two Types of Change in Storytelling

Two ways in which genomic knowledge changes or is incorporated in individual stories are presented by Mahr. First, we have a reaction that Mahr calls 'genetics as a relief'. As we learn from Mahr, the participants Kerstin, Nina, and Agathe embraced the genetic explanation of their disease. All three had had bad experiences with the psychosomatic explanatory model of IBD. From their perspective, the psychosomatic explanation implied that the illness was their own fault. The new genetic model is therefore, for them, a welcome and useful resource for their stories. They can use it to reshape the stories and to counter the implicit social blame connected to the old explanatory model. As Mahr puts it, the genetic explanation of IBD is used by some patients as 'a tool for self-empowerment against the social struggles connected with the older psychosomatic interpretation'. In this cases it becomes very clear that the meaning of the genetic explanation that it has for the patients lives originates partly, because of its potential to be used as a resource in existential storytelling and for changing the self-told story. Furthermore by changing the story the experience of the illness itself can be altered.

Interestingly, a second reaction to the genomic knowledge can be identified, which could be called 'denial' or 'opposition'. A group of participants is more sceptical about the classification of their condition as 'genetic'. For this second group, the genetic explanation entails relieving patients of their personal responsibility. So, in a way for patients such as Sabine, the genetic explanation does not fit into a coherent story about her life, because it potentially reduces self-responsibility and the motivation for self-care. Thus, when the genetic explanation is incorporated into the self-told story, a certain resistance or opposition can be noted.

As a result of Mahr's study, we see that the genetic explanation of IBD, provided by the new genomic knowledge, is experienced in very different ways. We can also see that the way in which the impact of genomic knowledge is experienced, depends heavily on the way it fits into the self-told story of patients, which in turn depends on many factors, including former experiences, the respective self-image, the social world, and fears and wishes for the future. The self-told stories are a way of coping with the existential crisis that is potentially induced by living with a chronic illness. As we have seen, genomic knowledge can be experienced as a useful resource or as a pushback.

But we also learn that there is a third option. Sometimes a new resource might not have much of an impact on the self-told story at all. The comparison with the heart disease case shows exactly this. Mahr reports that, in contrast to the IBD patients, genetic knowledge about the heart disease has almost no impact on the self-image, social world, or medical options of Elfriede's family. The knowledge of the health risk of cardiac infarction was already part of the family's 'tradition', and an integral part of their life-world. In a certain sense, the knowledge provided by a genetic explanation is not new for the members of the family but is only expressed in different terms. The genetic knowledge thus does not provide new resources for stories and it seems not to have much meaning for members of the family, though Mahr reports that their lives were strongly influenced by being part of a process of genetic knowledge production.

The construction of meaning by fitting things into a story is a fundamental (and some would claim primitive) means of explanation and understanding. As humans, we tell stories about ourselves to make sense of our experiences. The stories we tell can help us to cope with the existential questions raised by suffering from a chronic disease. Scientific understanding is usually thought to be a different kind of understanding. Science is supposed to find patterns that go beyond individual narratives. Wondering if there is any place for storytelling in science, Patrick Lewis (2011: 506) asks 'Where are the places and spaces for storytelling in research? Can research create a space for the storyteller and her story? Can a researcher become a storyteller? Can storytelling become research?' Mahr shows us very vividly that there is a place for storytelling in research, and furthermore how research has an influence on our self-told stories.

References

Arendt, H. (1958). *The Human Condition.* Chicago: University of Chicago Press.

d'Entreves, M. P. (2014). *Hannah Arendt. The Stanford Encyclopedia of Philosophy* (Edward N. Zalta, Ed., Winter 2016 ed.). https://plato.stanford.edu/archives/win2016/entries/arendt/. Accessed 9 August 2017.

Kenyon, G. M. (1996). The meaning/value of personal storytelling. In G. M. Kenyon, J. E. Birren, J.-E. Ruth, J. J. F. Schroots, & T. Svensson (Eds.), *Aging and Biography: Explorations in Adult Development* (pp. 21–38). New York: Springer.

Lewis, P. J. (2011). Storytelling as research/research as storytelling. *Qualitative Inquiry, 17,* 505–510.

Schechtman, M. (1996). *The Constitution of Selves.* Ithaca, NY: Cornell University Press.

Shoemaker, D. (2016). *Personal Identity and Ethics. The Stanford Encyclopedia of Philosophy* (Edward N. Zalta, Ed., Winter 2016 ed.). https://plato.stanford.edu/archives/win2016/entries/identity-ethics/. Accessed 9 August 2017.

15

Sudden Evolutionary Changes in Palearctic Warblers

Grischa Lichtenberger and Sarah Ambrosi

Images as supports of evidence (art installation at the Centre for Interdisciplinary Research, Bielefeld, Germany 2015)

warten, bis sich die sachen von selber auslösen. im traum dann kommen die bruchstücke wieder, vor denen man sich verstecken wollte.

The basis for the art work at the Centre for Interdisciplinary Research was reversal film slides that had been dumped by the University of Bielefeld, and which I had fished out of the garbage a few years ago. I was attracted to them by a kind of hermeticism. The fact that we no longer know for what kind of evidence they originally served, encloses the images in a space which, simultaneously, suggests meaning and conceals it. Were they explanatory some material for a lecture, or were they themselves the subject of experimental verification? Because I do not have any text (for them), all that remains is a residual trace of significance—that they should prove something, that they wanted to say something. Alienated from their context, even to the expert they can only say something

G. Lichtenberger (✉) · S. Ambrosi
Berlin, Germany

D. Mahr and M. von Arx (eds.), *De-Sequencing*, Health, Technology and Society,
https://doi.org/10.1007/978-981-15-7728-4_15

like: 'This is an image of a spectral analysis or a protein marker, etc.' The images lack some of the language whose mediators they originally were. As a former part of an argumentation, they themselves now lack the language to access the underpinning of which they were once made. The graphic attractiveness the pictures have for me is connected with a kind of residual demand within them—as they were still noticeably searching for their missing language.

In general, the reversal film slides extracted from the bulky waste of research are comparable to biographical narratives compiled from incomplete memories. Facing them, there is the feeling that one has always lacked language. This feeling is extended to a greater extent by the confrontation with linguistically charged images that seem to be older than the language written about them. And yet these memory images seem to have been bound to the subsequent language as soon as they were created. If one remembers pictures from childhood, one does not seem to have forgotten the language that belongs to them, but rather to have not yet found it. And at the same time, these reversal film images beg for linguistic integration—for grasping, retelling, understanding—which might be linked with the narrative potential that made these images enter into one's memory in the first place, just as the reversal film slides gained their graphic beauty by their lost integration into the argument of research, which turned them into silent speakers.

In even more general terms, it is a paradox of the image that (in its emergence) it wants to be a kind of cipher for the reconstruction of a non-narratable sensuality, but at the same time it retains only a ghostly trace of a symbolic quality in this sensuality, this substantial materiality to which it ultimately coagulates, for whose exhaustive decryption we ultimately lack the key. The image exists as such a marker, without ever being able to fully fulfil its function. Even with a pretended contextualization of the image as part of an argument, as was the case of these slides of research, there always remains a certain potential for resistance, a certain inaccessibility, a rest. It is precisely this remnant that allows the images to complete an argument, to prove it, to fill in a blank space in a logical connection.

Basically, I wanted to open a bracket with my work at the Centre for Interdisciplinary Research—starting from a material connected very concretely with the discourse of research and science. Its pictorial shape resembles the problem of the significance of informedness, which in my opinion is at the centre of the study of epigenetics. The biographical narrative stands for a memory image put into perspective by a noticeable restitution of historical, and partly even teleological, explanatory models of the informability of genetic material, which represents the increased interest in epigenetics.

In a second, parallel strand, this relationship of personal narration, pictorial metaphor and biological research is put into perspective in a different way. The subject here is the blackcap: for me, this bird is a kind of mythical animal. Where an outer motorway will soon be inaugurated (the A33, in Steinhagen, Gütersloh, Germany), there was once the house in which I grew up. The loudness of the soundscape from all the bird voices in spring and summer can hardly be described. And the few years in which I was able to visit my old home again from afar have intensified the experience connected to this location. The blackcap (Mönchsgrasmücke)—with its really concise, very varied singing, which is overpoweringly exponentiated by the small bird as if artificially potentised—was one of the illustrious visitors to the fruit trees in the garden.

earpiece, funnel: listen

A fragment of memory that accompanies me, always revives when I see the silhouette of a bird in the distance or hear a special song. It is the situation in which my father suddenly listens and searches the sky, or the bush or tree, from where he suspects is the source of the song. The outward-facing ear, the attention to the surroundings … how can I describe it?—It is in itself a kind of a metaphor, but seen through the eyes of a child it is also a very simple feeling of special love and admiration. It is a sign of hidden knowledge (about phenomena in the world 'out there'), a longing to be in the focus of this dissociation caused by these sounds? He then shows me the bird in front of the white Westphalian sky between the scissor-cut branches of the pear tree and names it: 'Look, there it is. Over there. A blackcap.'

[added with a whispering voice: what luck to have someone who says: 'Look over there'. cf: 'Dad, look over there!']

The Emlen funnel is a device used to collect comparative records on the migratory behaviour of birds. The test object, the bird, stands in a bowl of ink in the centre of a funnel covered with paper; above it is a net, an invisible boundary. When the bird begins (in vain) to flap its wings, it leaves traces on the edge of the paper-lined funnel with its ink-soaked feet, thus providing information about the direction in which it is trying to fly.

I have been living in Berlin for almost ten years now. The window where I write looks on to an inner courtyard. Opposite are a parking lot and a winter road maintenance depot. Behind these are the long street known as the Köpenicker Straße, factory buildings, then the River Spree. A few years ago, blackcaps nested nearby and could be heard occasionally in the inner courtyard.

The traces the bird leaves behind are not only evidence of its migratory behaviour, but also records of its desire to fly home. The paper lined funnel is like the walls of houses between which we retire, between which we wander like zoo animals in the cities, believing they are there to protect us. The invisible net above the bird's head is like the invisible barrier that makes it impossible for me to enter the memory from within the image function.

The speed with which migratory birds can adapt their breeding grounds to changing climatic conditions cannot be explained by a Darwinian selection by mutation. By transporting non-hatched eggs to the south from recently settled blackcaps further north, it could be proved that the artificially hatched offspring are still moving in the direction of the new home with the first generation (without, so to speak, having had empirical models for it).

The manufacturer's authorised representative in the EU is Springer Nature Customer Service Centre GmbH, Europaplatz 3, 69115 Heidelberg, Germany. If you have any concerns regarding our products, please contact ProductSafety@springernature.com

Printed and bound by CPI Group (UK) Ltd, Croydon, CR0 4YY
30/09/2026
02225286-0004